An Events Industry Takes Shape

This timely book critically evaluates the factors which shape an events industry as it develops, with the aim of helping to narrow the disparate behaviours and practices of organisers within the global marketplace of international events.

Stemming from an innovative qualitative research project, which included interviews with senior events organisers at landmark venues in both the UK and Poland, this volume provides an insight into both the emerging events industry in Poland and the developed events industry in the UK, highlighting cross-cultural risk and safety gaps that may impact organisers, clients, attendees, suppliers, and workers. The book highlights the importance of a unanimous global approach to events organisation, the creation of a professional community of practice, and ethos of self-learning within the events industry and the need for an international professional association for organisers involved with providing international events. The book explores the three themes of 'Event Culture', 'Tourism and Events', and 'Risk Awareness at Events', thus focusing on long-term factors of events industries.

International in scope, this book will appeal to students on courses such as managing events, planning events, project management, and hospitality and tourism studies, as well as event organisers in locations where events is an emerging industry.

Philip Berners is a Doctor of Philosophy in sociology, specialising in the field of international events. He is Associate Professor and Postgraduate Director at the Edge Hotel School, University of Essex, Colchester, UK. He is Fellow of the Higher Education Academy and a founding trustee of the Colchester Museums Development Foundation. Working in the UK events industry for ten years, Philip was head of events at landmark venues, including the London Hippodrome and Thorpe Park Resort. He has been the in-house event manager for corporations such as the Daily Mail Group. His client portfolio includes Her Majesty the late Queen Elizabeth II, the former Prince of Wales now King Charles III, the BRIT Awards, London Fashion Week, Formula 1, and global music stars such as Bon Jovi, Shania Twain, and Jennifer Lopez. Philip lived and worked in the events industry in Poland for ten years and founded an events academy in Warsaw. He also taught events management at Collegium Civitas University at the Palace of Culture and Science, Warsaw. Philip is the series editor for the Routledge Practical Guide to Events and Hotel Management Series.

An Events Industry Takes Shape

A Case Study of the UK and Poland

Philip Berners

Routledge
Taylor & Francis Group

LONDON AND NEW YORK

First published 2024
by Routledge
4 Park Square, Milton Park, Abingdon, Oxon OX14 4RN

and by Routledge
605 Third Avenue, New York, NY 10158

Routledge is an imprint of the Taylor & Francis Group, an informa business

British Library Cataloguing-in-Publication Data
A catalogue record for this book is available from the British Library

Library of Congress Cataloging-in-Publication Data
Names: Berners, Philip, author.
Title: An events industry takes shape : a case study of the UK and Poland /
    Philip Berners.
Other titles: Case study of the United Kingdom and Poland
Description: New York : Routledge, 2024. | Includes bibliographical references
    and index.
Identifiers: LCCN 2024004474 (print) | LCCN 2024004475 (ebook) |
    ISBN 9781032677538 (hbk) | ISBN 9781032684147 (pbk) |
    ISBN 9781032684161 (ebk)
Subjects: LCSH: Events tourism—Case studies. | Special events—Great
    Britain—Management. | Special events—Poland—Management. | Special
    events industry—Risk management.
Classification: LCC G156.5.E94 B47 2024 (print) | LCC G156.5.E94 (ebook) |
    DDC 910.68—dc23/eng/20240308
LC record available at https://lccn.loc.gov/2024004474
LC ebook record available at https://lccn.loc.gov/2024004475

ISBN: 978-1-032-67753-8 (hbk)
ISBN: 978-1-032-68414-7 (pbk)
ISBN: 978-1-032-68416-1 (ebk)

DOI: 10.4324/9781032684161

Typeset in Times New Roman
by Apex CoVantage, LLC

I dedicate this book to the memory of Ambassador Professor Ryszard Żółtaniecki (1951–2020): sociologist, poet, diplomat, and my valued friend.

Contents

Acknowledgements

The author acknowledges the following for their support with this project: Dr Gary Armstrong; Professor Eamonn Carrabine; Dr Nicole Ferdinand; Mr Olaf Olenski; Ms Małgorzata Tomorowicz-Malińska; and the organisers of events who generously participated in this project.

Thanks are also due to the team at Routledge, Harriet Cunningham and Emma Travis.

1 Introduction

1.1 Purpose of This Book

When a destination emerges into the global market for international events, the events industry in that country will have a different shape from a country which already has a long-established events industry. This is because a developed events industry has matured over much time and has had many opportunities to evolve its regulations and practices in response to event experiences. A mature industry will also have evolved a socio-professional construct for event organisers to network for knowledge sharing, such as professional associations. However, a country which emerges into the global market for events is thrust into the marketplace without having evolved the shape of its events industry. The difference in shape between one events industry and another is important because clients, attendees, sponsors, and workers are exposed to gaps in quality, service, and safety, depending on which country an event happens to be taking place.

An example of this is that one worker died during the build of the London 2012 Olympics – which might be considered an isolated incident, yet still tragic – but 'hundreds' of construction workers died for the Sochi 2014 Olympics (Shaw, Anin and Vdovii, 2015), and 6,500 people died during construction for the Qatar 2022 World Cup (*The Guardian*, no date). It can be seen in extremely serious terms, then, that there is a gap in the acceptance of risk between events held in the UK, Russia, and Qatar.

Exacerbating this problem is that stakeholders of events are largely unaware that there are gaps, what those gaps are, and what gaps are pertinent to which particular country. A worker at an event in Qatar, say, might unknowingly be closer to harm – or death – than if that worker were rigging an event in the UK.

It is not only workers who are exposed to gaps. Most organisers of events (clients) book venues abroad but might not be aware of the gaps in the country which they choose for their event. The gaps to which a client might be exposed could be in areas of quality, service, professionalism, knowledge, experience, or safety. And consumers who travel abroad to attend an event

DOI: 10.4324/9781032684161-1

(event tourists) are more prone to being unknowingly exposed to those gaps in quality, service, professionalism, knowledge, experience, or safety, depending upon where they choose to travel to attend an event.

This being so, little is known about the nature of an events industry, how it evolves, what factors shape its evolvement, and what impacts are felt by stakeholders due to differences between the shape of a young events industry and one which has matured. That is what this book is about: it reveals factors which influence the shape of an events industry as it develops and the causes of differences between an established shape and an emerging shape.

There are many areas where gaps in events might occur. However, the three themes in this book – Culture of Events, Tourism and Events, and Risk Awareness at Events – are derived from the author's insider knowledge as critical factors in shaping an events industry. These three key areas are central to managing events at international level, where it is crucial to be non-variable.

Event Culture defines the socio-professional cultural construct of the events industry in the two destinations that are the focus of this book: the UK and Poland.

Tourism and Events are inseparable when managing international events for foreign clients and travelling consumers – 'event tourists'.

Risk Awareness is vital across the global market for international events where gaps would expose event tourists to risk and harm.

There are other important elements of events that might have been considered such as technology or sustainability. But these are fast-changing and ever-moving in the marketplace and might not provide credible and consistent insights for the purpose of this book, which is to determine factors that shape an events industry. Technology, for instance, is variable depending upon the availability of equipment and specialist expertise, an event's budget, the complexity of an event and the resources required to produce that event, and the technical resources *in situ* at venues. Sustainability is becoming widespread but is a variable component dependent upon local laws as the driver, social attitudes, trends, and consumer-driven demand. This means elements such as technology and sustainability are unstable factors, whereas this book is concerned with factors of an events industry which must not be unstable, must not be variable, and must not have gaps. For it is the purpose of this book to highlight disparities in the global market for international events so that the gaps might get identified, recognised, understood, studied and researched, and, ultimately, narrowed.

The author has lived and worked in both the UK and Poland events industries for ten years, respectively. The events industry in Poland is in development, which is a stage of growth the UK events industry has already been through and emerged from. The different stages of development of events industries in different locations are why gaps appear within the global marketplace for international events. This is problematic because events management is an increasingly global phenomenon reaching emerging locations, which is evidenced by mega-events such as the Rio Olympics in 2016 (Davies, 2017), the 2014 World Cup in Rio De Janeiro (FIFA, 2014), and the 2012 Union of European Football Associations (UEFA) European football championship in Poland and Ukraine (UEFA, 2016).

The orientation for this book is in sociology – in particular, the socio-professional structure and social positioning of the events industry in the UK and in Poland – to understand the behaviours, approaches, and practices of event organisers in these two locations.

1.2 Book Structure

Chapter 1 underpins why the book and its outcomes have credibility because of the author's authority as a former producer of prestigious events in all genres for royalty, corporates, and global music stars. The author is now an academic with a PhD, a researcher, and textbook author.

Chapter 2 concerns the theme of 'Event Culture' to determine the socio-professional cultural construct of a developed events industry (the UK), and a developing events industry (Poland). The chapter contextualises factors which shape an events industry into a socio-professional construct of professional associations, events as a profession, formal qualifications, and operations such as after-event evaluation procedures, and setting objectives for every event, every time. At the conclusion of the chapter, the reader will understand why the shape of the socio-professional structure has influence on how organisers do their job, and the impacts upon their clients and attendees. Of note is how an events industry can manipulate its shape to narrow the gaps across the global marketplace of international events by constructing an environment where organisers learn from each other, their foreign clients, and their international attendees. Organisers in an emerging construct can learn from what has already happened in a developed construct which has been through experiences that have influenced its shape. Because of this, there is a role for tragedy in preventing an emerging industry from having to experience pain for themselves. Another factor of note is the need for organisers to be led to recognise their need to create for themselves an environment of self-learning where they interact with the socio-professional construct for acculturation learning in areas such as best practice, safety, training, accreditation, and learning about new legislations.

Chapter 3 presents the second theme of 'Tourism and Events', which is an essential component of international events at landmark venues in capital cities, which, in this book, is London and Warsaw. The chapter explores event tourism, the trend of tourism, penetrating foreign markets, the internationalisation of events, and the experience economy. Of note is the difference between the developed and the emerging construct in their approaches to foreign business. The former is dynamic and proactive with a plurality of approaches to win foreign business; the latter is passive, non-interventionalist and waits for foreign business to arrive. Also covered in the chapter is the impact on events in hotels when an events industry takes shape, and why hotels lose events business. This leads to a section on gaps in service quality being accepted as a lower tier of event quality and safety because of practices becoming embedded in the socio-professional culture at that location.

Chapter 4 examines the theme of 'Risk Awareness at Events', which is an essential theme for every event organiser, everywhere. The chapter addresses the management of risk at events, how organisers are learning about risk, variances in levels of accepting risk, organisers' attitude to risk, learning from experience, the self-learning environment, and considering the risks to consumers (consumer risk perspective). Although it is identified in the chapter that there are omissions in risk control procedures by organisers in both the developed and developing constructs, the reader will understand why organisers fail to manage risk, why organisers in different locations have variant levels of risk acceptance, and why clients and attendees are closer to harm at an event happening in one location over another.

Lastly, there are notes from the author on the outcomes of this book.

1.3 Context of This Book

This book stems from an original research project described by Dr Nicole Ferdinand of Oxford Brookes University as 'one of the extremely rare cross-cultural studies of the event industry' and 'a worthwhile contribution to [the] field'. Dr Gary Armstrong of City, University of London, defined the original study as a 'most unusual topic [which] provides for a much-needed insight into how various large-scale leisure events are both considered and delivered'.

The UK is selected as representative of an events industry at an established stage because it is one of the world's most developed event industries in terms of destination infrastructures, venues, and service suppliers, with 1.3 million events in 2017 in the 'business' sector alone. The Business Visits and Events Partnership (BVEP) (Business Visits and Events Partnership, 2020) estimates that the UK events industry is worth £70 billion and accounts for over 50 percent of the UK visitor economy. The UK events industry employs over 700,000 people in the sector (BVEP, 2020) and is ranked fifth in the top 20 cities ranked by the number of meetings organised in 2017 (Conference & Incentive Travel, 2018), after Barcelona, Paris, Vienna, and Berlin. Selection of the UK as a developed construct is further justified by the turnover of £340.1 million by the top ten UK event agencies (Conference & Incentive Travel, 2018) and the rise in applications for event management courses (AEME, 2008, cited in Goldblatt, 2014). London was ranked as the top EMEA (Europe, Middle East, Africa) destination for meetings and event planners in 2018 (*Conference News*, 2019). Still, the UK events potential is continuing to evolve and is not stagnant, which is indicated by a submitted planning application to construct a £200 million global event campus in the Glasgow area (Wood, 2018). Within the UK events industry construct, this book focuses on London as the comparative capital city to Warsaw.

Poland is selected as representative of an emerging and developing events industry with the meetings industry contributing an estimated 1.5 percent to gross domestic product, employing 220,000 in the meetings sector, and 2,031,589 consumers attending events (Buczak *et al.*, 2020). The Poland events industry is in rapid growth stage, with the International Congress and Convention Association reporting an increase in the number of meetings held in Poland (with the criteria of lasting at least three days and with at least 300 consumers of which at least 40 percent are foreign) as increasing from 150 in 2009 to 230 in 2019 (Buczak *et al.*, 2020). The Poland Convention Bureau reports that just 15 percent of corporate events in Poland are of foreign group origin compared to 85 percent domestic, and 9 percent of events are hosted in congress centres and 16 percent in event venues, compared to 65 percent hosted in hotels (Buczak *et al.*, 2020). The same report shows that trade exhibitions are just 7 percent from foreign origin compared to 93 percent domestic, demonstrating how Poland is not yet penetrating foreign markets for international event clients and events which attract foreign event consumers. Although Poland has already emerged into the global marketplace for international events, the 'origin' (Buczak *et al.*, 2020) statistics demonstrate that Poland has yet to increase the level of events business available to draw from foreign markets. Warsaw makes for a comparative capital city to London.

1.3.1 Acknowledging Poland's Post-Communist Society

While this book is focused on the socio-professional cultural construct of an events industry taking shape, it must be acknowledged that the events industry in Poland exists in the cultural and societal contexts of a communist legacy, in contrast with the UK events industry having evolved in a long-standing democratic society. This does impact the shape of the two comparative event industries but not in a way that is detrimental to the outcomes presented in this book. In fact, this difference strengthens the need for understanding the causes of gaps in all locations across the global market for international events. Poland is one of the fastest-growing economies worldwide and the fastest-growing economy in Europe and was the only country in the European

Union (EU) to avoid recession during the financial crisis of 2008 (Bogdan *et al.*, 2015). Since the 1990s there has been enhanced competitiveness of Polish industry and narrowing of the gap between Poland and the EU, which is driving Poland closer to the standards and structures of other EU countries (Domański, 2003). These factors fit well with the cross-cultural comparative study of an emerging events industry and a developed events industry, identifying differences in the behaviour and practices of the two sectors, and what gaps exist and how they might be narrowed.

In 1989, Poland was the first country to leave communism peacefully and form the world's first post-communist government (Kemp-Welch, 2008). The fall of socialist ideology brought rapid and profound social change and development, which continues to have long-lasting impacts on social class, social change, modernity, globalisation, and identity (Outhwaite and Ray, 2005). The evolution of a post-communist society has been described as democratisation backwards, suggesting the search to identify with democratic alliances such as the EU, which Poland joined in 2004 (Lane and Wolanski, 2009). This is further reason why planned events provide a suitable and appropriate framework for this book, because it is a visible and obvious evolution of an industry identifying with the global market. In doing so, Poland has rapidly emerged onto the world stage for international events by hosting significant international events, including the UEFA European Football Championship in 2012 and the COP24 United Nations Climate Change Conference in 2018, but also in promoting Poland as a destination for event tourist clients and consumers.

The disparity in the fabric of society between the UK and Poland will have bearing upon the outcomes presented in this book. For one thing, Howard (2006) observes that the Polish state had not provided the resources and support for the organisation of a civil society, which contrasts with the UK. But in modern Poland there are resemblances to a constitutional system, because the cornerstone of reform in Poland was the Local Self-Government Act, which drew heavily on the Council of Europe's European Charter for Local Self-Government (Howard, 2006).

Communist versus democratic societal differences aside, this book does squarely meet its objective of identifying how emerging event industries take shape in the context of the global market for international events. It is the socio-professional cultural construct of an events industry in the context of its location which is the focus of this book, not the wider sociological or cultural implications of any particular society. Every location will have its peculiarities and idiosyncrasies fashioned from its society, history, and cultural development, regardless of whether that location is post-communist or not, which is why the outcomes presented in this book relate to any emerging industry, anywhere.

1.4 Background of the Original Study

This book is based on an original study by the author to determine factors which shape the development of an events industry by identifying differences in the behaviours and practices of organisers in an emerging events industry with one that is already established.

The study objective was to research event organisers in both the UK and Poland within the socio-professional context of the events industry in which they live and work.

Researchers agree that the global events market is a transient joined-up sociocultural community. But there is a void of cross-cultural research into the disparities between event locations at different stages of development. The differences not only are cultural but raise sociological implications in the management of events in different locations and the variance of exposure to which stakeholders (clients, consumers, suppliers, sponsors, and workers) are exposed.

1.5 Credibility of This Book

Twenty-four organisers were interviewed and observed as actors in their respective events industry and it is important to interpret the way they act in the UK and in Poland to draw similarities and/or differences. The original study utilises empiricism, which demands the capture of tangible evidence to substantiate interpretations and assumptions to acquire knowledge from the collection of observational data, and so it is fact-based rather than personal opinion or speculative imaginings (King and Horrocks, 2012). Empiricism facilitates reliance on the stories the interviewees are telling, which, although may be partial, is adequate for purpose (Becker, 2007). This means that the interpretations of the data are not derived from perspective but are grounded in the responses of 24 interviewees.

It is not easy to separate interpretations from fact (Becker, 2007), and a fact in its social context implies and invites interpretation. It is therefore important to adopt interpretivism to interpret the stories being told during the interviews and to support autoethnographic insights. Veal (2011) points out that interpretivism involves participants providing their explanations of their situations or behaviours for the researcher to see the world from their point of view, which is the intention to subdue any presuppositions. Being an insider on the subject and knowing industry practices, jargon, and behaviours is helpful to interpret the actors' storytelling during the in-depth interviews, led by the responses of the interviewees which in fact determines repetitive patterns of behaviours and practices across the population.

1.5.1 Interviews

Twenty-four in-depth interviews are conducted with senior-level organisers at some of the best landmark venues in the UK and Poland. Seventeen interviews are with UK organisers, and seven with Poland organisers, which elicits valuable and credible data and fulfils the requirements and objectives for the basis of this book. The data from the interviews show signs of repetition and saturation, leading to identifying patterns for data interpretation.

Mason (2014) acknowledges that while qualitative interviewing involves what may seem to be a relatively small sample, it is made up for in the very rich data gathered. It is relevant to retain quality research over quantity research, which is underpinned by Muncey (2010) confirming that in the social sciences a large sample without rigorous control of variables produces inferior quality evidence. The interviews are further supported and underpinned with observations and autoethnography, which is a justified research approach when the sample involves small numbers of people (Muncey, 2010).

Semi-structured questions are used in the interviews to follow a set theme with each participant for consistency of analysis and comparison, but with the flexibility to vary the order of questions and ask further questions (Saunders, Lewis and Thornhill, 2016). Structured questions would restrict participants from exploring the themes being presented to them. Unstructured questions would not work either, because of the predetermined list of questions for comparison and consistency (Saunders, Lewis and Thornhill, 2016; Mason, 2014) across the two locations of the UK and Poland.

Interviews are recognised by Qu and Dumay (2011) as a way for researchers to learn about the world of others, which is the aim of the primary research activity for this project. Insider knowledge of the events industry is helpful for quickly establishing a rapport with each interviewee. None of the interviewed organisers are familiar with providing their insights for a book and so could hold some trepidation of what is expected from them, what would happen, how the information they offer would be used, and would what they say be usable. Each interview does, in fact, provide rich and usable data.

Knowledge of the Poland language allows the interviewer to insert Polish words to bond with the interviewee. Use of the language also allows interpretations on the intonations and phraseology when Poland interviewees are speaking English, and there are no identified barriers or misinterpretations due to language differences.

The 24 senior-level event organisers at landmark venues in the UK and Poland can be considered as 'representations of society' (Becker, 2007, p. 5).

There are classical definitions of what constitutes a landmark, including this from *Merriam-Webster* (2021): *A conspicuous object on land that marks a locality, or a structure such as a building of unusual historical and usually aesthetic interest, especially one that is officially designated and set aside for preservation.*

Closer to the recognition of a landmark venue for the purpose of hosting an event can be found in the *Cambridge English Dictionary* (2024): *[A] building or place that is easily recognised, especially one that you can use to judge where you are.*

These definitions help underpin justification that the venues identified for this project could be categorised as landmark. This is important for the use of like-for-like venues in both the UK and Poland, and by doing so, it defines the type of interviewee, their types of consumers, and their clients because typically a landmark venue is in a prominent location, it is large, it is expensive to hire, and it will be known to consumers and clients.

Although each landmark venue owns unique peculiarities (Tonnelat, 2009) – which is what makes it a 'landmark' venue because it is not replicated anywhere else – the selecting of the right type of venue needs to fit with the three themes of this book (Event Culture, Tourism and Events, and Risk Awareness at Events).

However, there are no definitions, classifications, or categorisations of landmark venues in current and available event literature, so a set of criteria for target venues is created and adopted. First, to determine what constitutes a 'venue' reflects the social practices of attending an event because the interpretation of what is a venue resides in social practice and not just 'in the heads of individuals' (Dey, 1993, p. 11). For those attending a venue, they require further social practices such as getting to the venue, buying a ticket, and audience behaviour (Sayer, 1992, cited in Dey, 1993). The culmination of the social practices and social construction in putting on an event for people to attend happens at the place where there is a live happening: the venue (Berners, 2013, 2017, 2019). Thus, the set of characteristics for the sample of venues that would meet the project requirements is

> [a] building of significance that is known for hosting events (is an event venue); is built for the purpose of hosting events; is widely recognised in its location and is therefore easy to locate; and has a reputation for hosting large events (capacity upwards of 200 attendees).

In London, this includes *Olympia* exhibition centre, which hosts over 200 events each year with a maximum capacity of 7,000 consumers (*Olympia*

London, 2018); *ExCeL* convention centre, with over 400 events each year and over 4 million consumers annually (*Visiting ExCeL*, 2021); and The *o2 Arena* hosting over 260 events each year, with a capacity of 20,000 (*The o2*, 2021). In Warsaw, it includes the *National Stadium of Poland* (*Stadion Narodowy*), with a capacity of over 58,000 (*PGE Narodowy – Warsaw National Stadium – The Stadium Guide*, no date); the *Palace of Culture and Science* (*Pałac Kultury i Nauki*/PKiN), which is an iconic 44-storey building in central Warsaw with ballroom, congress, and banqueting facilities and a capacity of 4,500 (*Pałac Kultury i Nauki*, 2021); and *The Royal Castle* (*Zamek Krolewski*) in Warsaw Old Town, which is a former royal residence of significant historical importance dating to 1598 and hosting a range of events from banquets, dinners, fashion events, and hosting dignitaries (*zamek-krolewski. pl*, 2021).

Although all 24 interviewed organisers are white (13 male/11 female), which raises a socio-demographic question, the selection grounds the objective to identify gaps between the UK and Poland events industries. Identifying interviewees is driven by, first, the criteria of the venue meeting the set criteria outlined earlier. Once the venue is identified as representative, it is the venue which leads, second, to the organiser at that venue. This two-step approach reinforces that the interviewee is legitimate for the purpose of the study and indeed employed at a high level of organising events and in a landmark venue.

For anonymity purposes, there is a code for each of the interviewees. The anonymity code depicts within brackets the country of the interviewee, followed by the number of the participant, and a generic description of their type of venue. So, the label for Participant One in the UK who is an organiser at The o2 Arena is: '(UK1 – landmark arena venue)'. Where a participant is not venue-based, the generic description is of their role, so the label for Participant Four in Poland who is an organiser of charity events is: '(PL4 – organiser of charity events)'.

1.5.2 Observation

Observation is a hallmark social research method for capturing unstructured qualitative data which, for this project, is not for observing the job of organisers and what they are doing but to substantiate the story they are telling (Dey, 1993). Plummer (2019) acknowledges that stories are weaved and interconnected with truth and fiction as determined by the storyteller. Observation is a tool for when each interview is concluded, to view the venue, for one, and extend the opportunity for data capture in a more social and free setting. Besides, most organisers enjoy every opportunity to show their venue to a visitor.

Governmental travel restrictions during the COVID-19 pandemic halted in-person interviews and the observation of venues. So, video-call interviews are used and 'retro-observation' drawing on recollections of each venue from when the author was an event practitioner in both the UK and Poland.

The rationale for applying observation to corroborate what is being told in the interviews proves justified. For example, PL1 provides 'straight A' answers which indicate their venue is excellent and without faults. However, observing a significant concert at that venue – which was the Warsaw leg of a worldwide tour – revealed unused chairs stacked in an emergency evacuation route, which is a fundamental error and breach of fire precaution laws.

1.5.3 Autoethnography

Autoethnography is a method which facilitates the use of studying the experiences in one's own life as a means of illuminating more general social processes (Hammersley and Atkinson, 2007). Chang (2008, p. 13) describes the concept of culture as a 'web of self and others' and that the autoethnography research method utilises the researchers' autobiographical data to analyse and interpret their cultural assumptions. Adopting autoethnography, then, utilises the author's significant wealth of knowledge from a time when in the field.

It should be noted that autoethnography is a method personal to the researcher: autoethnography or 'personal ethnography' is a 'first-person method' of research (Desjardins *et al.*, 2021, p. 37:4), where the researcher's first-hand experience is the starting point of knowledge production and can involve observing one's own experience with a particular group or subculture. This is a first-hand study of society and culture in action (Murchison, 2010) but with the application of a biographical research method (Chang, 2008). Desjardins *et al.* (2021) point out that simply going into the field to write about it and the people being observed can be a distanced perspective, whereas employing autoethnography allows the researcher to input their own voice in their writing and provide an authorial voice in describing patterns.

This book, then, is written from perspective and experiences which would be different from somebody with a different ethnicity, gender, and career profile. This bears influence on the outcomes, but they remain valid and representative because of the organisers and their perspectives, and because the interpretations are grounded with existing literature as well as empirical data.

1.5.4 Validity and Reliability

From an epistemological viewpoint, interpretation of the stories being told (Becker, 2007) during each interview could lead to subjectiveness. However, Dey (1993) attributes that the difference between interpretation and misinterpretation is in the reluctance to admit the possibility of misinterpreting data. Lakshmi and Mohideen (2012) point out that validity in research requires multiple sources of evidence which act as a representative sample, and that reliability is the degree of consistency in the measurement procedures leading to results. To meet these criteria, an acceptable level of interviews is conducted and triangulated with retrospective observation technique and autoethnography to identify patterns from the data and interpret valid and reliable findings to present the outcomes.

2 Event Culture

This chapter defines the socio-professional cultural construct of the events industry in the UK and in Poland. This is important in positioning where events sit in both a social and a cultural context of professional society in the two locations.

The UK is further ahead with events than Poland, so the aim is to understand what factors influenced the UK events industry as it took shape and whether this is potentially helpful to shape the emergent Poland events industry. To be explicit, this book is not exploring or analysing wider cultural complexities outside the socio-professional construct of the events profession, such as religion, gender, clientelism, command cultures, or long-time democratic versus post-communist nations.

2.1 Cultural Context

2.1.1 Professional Associations

The socio-professional construct of the UK events industry is overcrowded with 42 industry associations (Bowdin *et al.*, 2011, p. 31), which has the effect of causing confusion, such as which is the best association, why is one association different to another, and which association is a credible accreditation for its members. An events practitioner in the UK will find it difficult to know which or how many professional associations to join, or why they should join any. It is easier not to join and get on with the job of organising events to the best of their ability using the practical knowledge and experience they gain along the way. The existence of many associations fragments the unity of collaboration, which is counterproductive to the purpose of industry associations. It is a point realised by Bowdin *et al.* (2011), who note the need for consolidation to happen for the industry to move forward. It seems, then, that as the events industry in the UK developed, it has impeded itself.

DOI: 10.4324/9781032684161-2

Nonetheless, UK organisers are benefitting from participating with professional associations. They gain training, knowledge of industry best practices, networking opportunities, and knowledge sharing. They also gain accreditation and compliance from some of the industry bodies such as National Examination Board in Occupational Safety and Health (NEBOSH) and PRojects IN Controlled Environments (PRINCE2).

The problem is that there are so many professional associations in the UK for an organiser to participate – they will not know them all and cannot participate with each. It thus becomes a lottery of knowing which to join, if any, and is off-putting.

In contrast, with the few associations already emerging in Poland, such as the Poland Convention Bureau, and Poland Meetings, Incentives, Conferences, Exhibitions (MICE), organisers in Poland are blissfully not being bombarded with professional associations vying for membership sign-up with promises of benefits and teaching the right way of doing things. It is easier for organisers in Poland to navigate their way to belonging to industry bodies and can thus bind the industry. If the few associations are wise, they will absorb more organiser-members as the industry expands and professionalises, rather than establish more associations. This is certainly good for the associations and will continue holding the industry together for knowledge sharing and networking within the few industry networks.

While their events industry forms its shape, Poland can learn from the glut of professional associations in the UK events industry. In its developmental stages, Poland could learn that if the number of associations grows with the growth of the industry, it will splinter the cohesion of the events industry as has happened in the UK. But it will prove difficult for Poland to curtail the growth of associations. First, it would require legislation to restrict associations from springing up, which is unlikely to happen because it cannot be considered a legal problem requiring the protection of law. Also, restrictions by the State on what would be considered as the growth of an industry will be unpopular and would recall the former state of communism. Second, growth is organic and changeable, so who is to determine how many associations should there be to support the industry, and how many is enough, and how many is too many. Poland, therefore, is unlikely to avoid the disadvantages of oversaturation of professional associations in the socio-professional construct of their events industry until it becomes unsustainable and hinders the cohesion of their events industry. By this time it is impossible to reverse.

In the meantime, before Poland reaches that level of overabundance of professional associations, there are too few and their events industry is taking shape without a socio-professional construct of support, learning, and knowledge transfer and is driving an autonomous culture in the Poland profession of organising events.

What is required is a balance between just enough associations to cohere and professionalise the profession but not too many that will confuse and fragment the industry to the point of anti-coherence which is the antithesis of the objective of professional associations. The problem is that the development of an industry is organic and ever-changing so will always be difficult to measure the 'right' level of professional associations. Nevertheless, it is helpful to be aware of the implications of oversaturation.

Regardless of the above arguments as to how many professional associations should there be, they will serve some purpose to organisers of events. Even though the UK events industry has reached the point of being awash with associations, UK organisers evidence benefits from engaging with some of the range of industry bodies available to them:

> *The typical ones for us would be the Association of Event Venues [AEV] and the Association of Event Organisers [AEO] . . . Sharing best practice and helping to elevate the standards of our industry. We'd be in competition with other venues but actually the other side of it as a group of venues across the country it's our job is to collectively improve our standards and offer our customers the best that they can possibly get.*
>
> (UK4 – landmark convention centre)

This organiser recognises the value of collaboration, with the goal of furthering industry best practice. They place value in collaboration above the perceived threat or fear of competition. Socialising with like-minded professionals has the effect to dissipate the fear of competition because organisers will realise there is enough business to go around and that other organisers actually pursue a wide range of event genres from other sources and other markets.

Even when organisers are in direct competition – such as within the limited socio-professional construct of an emerging location – they can learn from each other and get better (the self-learning paradigm) through competitive advantage, and this should be encouraged to happen. So, the threat or fear of competition is outweighed by the need to self-develop through participating in the socio-professional construct of the events industry whether it is developed or emerging.

The impetus for education and professionalism in an events industry is driven by an individual organiser's recognition for self-improvement of their performance, rather than by the demands of consumers, or even the insurance industry responding to past tragedies. However, the need to comply with legislation must be a motivating factor for the individual organiser or the company they work for, so it is reasonable to state that the law is in fact the key driver for professionalism. This is unsurprising because, so far, the events industry has abstained from requiring or even acknowledging the merits of professional education or qualifications.

Until events management as an education discipline evolved in the UK, nobody in the events industry was interested in an organiser's education or qualifications. This is why some firms and individuals are trusted implicitly in this industry because of their reputation. Experience builds knowledge, and so an organiser becomes trusted based on what they have done before. The quality and profile of events which an organiser has produced, or the reputation of their venue, outweigh any qualifications. In an emerging events industry – as is the case in Poland – specialist formal education is not valued.

The previous excerpt from an interviewee shows that this organiser is keen to share best practice and participate with elevating the standards of their industry. This is important because knowledge, and how knowledge is identified, collated, and shared, is at the centre of events management (Brown and Stokes, 2021). This organiser does not want to work autonomously within the limited parameters of their own venue but takes a holistic view of the industry to which they belong.

Events in the UK is a fiercely competitive industry to win clients, because there are many players in the market, and venues are in direct competition with each other – particularly in London where there is a greater range of landmark venues from which to choose. It is surprising, therefore, that an organiser in London would not be concerned about competition but show their willingness to work as a wider collective in order to maintain and enhance industry knowledge and best practice. This might be true of an organiser at a secure landmark venue which enjoys high-volume and high-spend business levels. An organiser at a smaller or independent venue might not be as open to sharing in the competitive marketplace. This is a barrier to networking within the socio-professional structure of both a developed and an emerging industry.

Nevertheless, there is evidence in the UK events industry of a community of practice (Brown and Stokes, 2021) in action which connects event professionals to a community with modes of working, shared values, and practices. This helps in amalgamating the industry and will go some way to forming standardised practices albeit at a local (domestic) level. Becker (1984) categorises such communities of collaboration as an 'art world' consisting of people whose activities are needed for the production of art, and that these same people would co-operate as an established network of links among those engaging within the field for benefits such as networking and knowledge sharing: 'If everyone whose work contributes to the finished artwork does not do his part, the work will come out differently' (Becker, 1984, p. xxiv).

All organisers rely on an art world because the nature of organising events requires the procurement of resources from a diverse range of specialisms such as technical, catering, floristry, decoration, safety, and security. Even a highly experienced organiser who is proficient in their role cannot be a specialist in all areas nor perform them all. But it is their role to bring all the elements together and oversee them for the purpose of a successful event.

Poland has not yet evolved a socio-professional network of industry-related professional associations important for supporting event organisers and providing a culture of collaboration instead of competition. This difference is not surprising between a developed industry and one that is in its forming stages. However, it reveals a gap in the wider global market for international events because the absence of the opportunity for organisers to network within a supportive and collaborative environment leaves them at a disadvantage and behind a developed event industry:

> *On all these forums they arrange staff training days . . . the advice is free and you can share the issues. When you look at a problem you've got, you're probably not the first person to have that problem and other people will have experience of it. I think it's learning from the best practice of others, having benchmarking groups where you can share information, and learning from others' successes and failures.*
>
> (UK5 – organiser of local authority events)

This statement is spoken from the perspective of an organiser who is already participating with professional associations and reflecting on the benefits they are receiving. Poland organisers are not receiving such benefits because of the absence of professional associations in the limited socio-professional construct at that location.

It raises the implication of how organisers in an emerging industry are developing their way of organising events if they are doing so naively and autonomously. The previously mentioned point about competition is likely a factor because there are fewer landmark venues in the location of the emerging market. Thus, competition is fierce and organisers will be less willing to engage with a wider socio-professional network, even if it were available to them. Perceived competition is a barrier which could be mitigated through closer-aligned working relationships across the socio-professional construct, which professional associations do facilitate.

Poland organisers are not exempt from rapid cultural changes, and they do need to be part of a socio-professional network to identify and understand trends and changes in the events marketplace. In both the UK and Poland, the uptake of social media, updated regulations and legislations, advances in technology, and new trends such as sustainability and dietary awareness are driving consumer demand:

> *People are much more aware of sustainability and also wellbeing. Wellbeing and sustainability are the two key things that people keep asking for at the moment. In terms of how events have changed, that comes into it because people are looking at sustainability so that they don't have to fly thousands of miles to get groups of people over from one end of the thing to the other . . . That's what the industry basically says as well in terms*

*of a lot of the people I've spoken to over the last six months or so. . . .
Current trends are sustainability. We have a lot of CSR policies as well
in place so that we can show people we do community engagement and
charity work, as well. In addition to that, wellbeing is a huge talking point
at the moment. We have an entire floor of the hotel, level 4, dedicated to
wellbeing. Previous to that it was all about rooftop gardens and private
dining rooms. They still come up but the two key trends are sustainability
and wellbeing.*

(UK13 – sales account director)

It can be seen from this organiser how important it is to win business by
promoting trends as added-value amenities. It requires organisers having
to very quickly and constantly adapt their offer to meet changing client and
consumer trends, which are recognised from networking in the industry.
Lenski, Nolan and Lenski (2014) refer to change as 'socio-cultural evolution' resulting from a society gaining new information and new technology.
Organisers in an emerging environment must anticipate trends or identify
them but this will be difficult if performing in a shape without a socio-
professional construct. Reacting to trends already taking place is too late
because it takes time to pivot services and the trend could already have
changed. This means that an emerging events industry will always be trying to catch up or, worse, always be behind the curve of trends. Besides, a
newly shaping events industry needs to remain sustainable and continue to
emerge until stable and, eventually, developed. Thus, competition cannot be
considered as a barrier to socio-professional networks and engagement but
should instead be seen as a supportive and collaborative art world or community of practice for the collective events industry, and the actors within
it, to win business.

There are benefits from engaging with industry associations and the
socio-professional network because it influences the way organisers do their
job if they are exposed to professional elements such as knowledge sharing
and knowledge transfer. Hanquinet and Savage (2016) see this as infusing
social actions with culture through representations, narratives, and structures
informing, regulating, and shaping social actions. Organisers in a young
events industry are bereft of this influence but it can be seen in action in the
developed community of practice in the UK with organisers benefitting from
networking, communication and liaison, training, codes of ethical practice,
and lobbying on behalf of their members (Bowdin *et al.*, 2011):

*Being part of international associations we have exposure to meetings,
conferences, seminars . . . we do a lot of work with the AEV* [Association
of Event Venues]*, I chair the event management committee of that association and we talk a lot about working with international clients and how we*

as UK venues set best practice, especially safety, how we set standards of best practice.

(UK6 – landmark convention centre)

This raises another dimension for participating with industry associations because some organisers view it as important to be associated with professional associations for reasons other than networking and knowledge sharing, but for accreditation purposes. This seems particularly true with Poland organisers in the shaping industry who seek accreditation from belonging to the few industry-related associations that do exist such as the Poland Convention Bureau (*Poland Convention Bureau*, 2020) and Polish MICE (*MICE Meetings Incentives Conferences Events Incoming Tour Operator*, 2018). Poland organisers seek to belong to the narrow range of professional associations for accreditation of their services to be recognised as bona fide. It is accreditation, rather than seeking a learning environment, which is driving organisers in Poland to join a professional association. Although, learning is a benefit which may become realised after joining. The motivation to join a professional association in the Poland events industry, then, is foremost for accreditation and/or marketing purposes.

This is helpful with the professionalisation of the emerging industry because organisers who participate with professional bodies – for whatever reason – will learn the regulations and legislations to which they must comply. Such regulations and legislations are brought about from previous disasters at events, so the industry is in fact shaping itself around the learning from previous mistakes. It is the law, then, which drives organisers to professionalise their behaviours and practices.

The problem with accreditation is that it does not mean much to anybody outside the industry's socio-professional network, such as consumers and personal clients who likely do not know what the accreditation body is or does. Although it is true that corporate clients and MICE clients will be aware of some of the accreditation bodies, so if that is the organiser's client base, it would be helpful to display those accreditations. In general, however, most attendees and many clients will not understand why the organiser is accredited, and it is not a qualification or membership of a required institution such as a university teacher being accredited by the Higher Education Academy, say, so it carries limited value.

There are benefits for organisers being accredited to some of the industry affiliations and doing so might mean personal and professional development and a requirement to uphold a level of standards, which is a good thing. But when observing the developing Poland events industry, affiliated accreditations tend to be primarily for marketing purposes, rather than for the purpose of knowledge gain and transfer. In this case, they will not make much headway in bringing Poland organisers closer to the standardisation of event management practices or narrow the gaps between events in Poland and other locations within the international events community, which is a lost opportunity.

Reflecting towards the hospitality sector – which remains a closely related sector to events – there are accreditations which have reached recognition by consumers: for example, the Michelin Star or the AA quality rating systems. The global events industry could achieve this as well if it managed to establish an international accreditation not only for industry professionals but also for consumers (clients and attendees) to recognise excellence in areas such as qualifications, training, experience, safety, quality, service, and standards.

2.1.2 Events as a Profession

The presence of industry associations professionalises an industry, although Klegon (1978) questions the definitional criteria by which professions can be distinguished from non-professions. Macionis and Plummer (1998, p. 435) categorise 'profession' as 'work based on theoretical knowledge, occupational autonomy, authority over clients and a claim to serving the community'. But Klegon (1978) persists that for this approach to be viable a consistent set of traits of professions need applying with limited ambiguity. Perhaps this is why Pierre Bourdieu uses the concept of a 'field of forces' (Harker, Mahar and Wilkes, 1990, p. 8), which he developed after personal experiences 'in the field' in order to examine the social space in which interactions, transactions, and events occur (Grenfell, 2012).

The discourse, however, has moved from the 'sociology of professions' to the 'sociology of professional knowledge' because of the need to distinguish professions from other occupations and the need to deploy expert knowledge in minor occupations and semi-professions (Young and Muller, 2014). Yet Eyal and Pok (no date) argue the case for the 'sociology of expertise' because profession is about fields and jurisdictions whereas expertise considers spaces between fields. Whatever it gets called, each recognises how expert knowledge distinguishes professions from other occupations and that all professions involve some practical expertise, which leads to considering knowledge as a qualifier of expert occupations. Applying this to the culture of the events industry in the UK identifies that the UK does have a professional industry, and so does Poland – it is just that each are at different stages of development:

Developed through knowledge and experience.
(UK7 – landmark conference centre)

[No qualifications] *but practical experience.*
(PL2 – landmark venue)

[My] *formal education was within the process of preparing for Euro2012 because within the company policy we had some project management – it wasn't event management but project management. Apart from this it was through experience mainly. More experience and less formal education.*
(PL6 – landmark convention centre)

These excerpts, and others from the interviewed organisers, demonstrate how organisers in both the developed and developing constructs value their knowledge above their need to study for an events qualification. This mindset determines the direction of how the culture of an event industry takes shape without the presence of a community of practice to evolve the learning of how to do the job, which then becomes embedded as the normal way of doing things.

It is characteristic of Bourdieu's 'habitus' concept (Grenfell, 2012), whereby social practices are characterised by the occurrence of regularities without the presence of rules to dictate practices. This was observed by Marciszewska (no date, cited in Smith and Robinson, 2009) that a knowledge-based economy is emerging in Poland to replace the previously dominant national culture, which she attributes to the impact of intense competition brought about by globalisation in the sectors of culture and tourism. Douglas and Wildavsky (1982) alluded to this as well by informing that we owe our current perspective to modernisation.

The concern is that if knowledge has developed through Poland's way of doing things, it does not make it the right way of doing things, although Foucault recognises that strange practices are actually governed by structural codes of knowledge (Carrabine, 2007). Perhaps it is just different knowledge and might even be skewed by sociocultural factors such as emerging from a communist regime.

As an industry professionalises, it requires a culture of regulation and education in both the '*re*-production' of knowledge and experience from one generation to the next and '*pro*-duction' through the generation of new knowledge (Bourdieu, 1973, cited in Harker, Mahar and Wilkes, 1990). Until recently, however, the events industry in the UK has eschewed formal education or qualifications in events management. This is mainly due to the relative newness of the discipline, because until recently there was no provision of university courses in events management.

The recognition of professionalism for an organiser, then, has been driven by testimonials, reputation, networking, client portfolios, and job history – not by their education or qualifications. In fact, this is why some firms and individuals are trusted implicitly and their word and abilities are considered as more significant than any qualifications.

Trust is all-powerful in winning event jobs because a client is making the decision to place the success of their event in the hands of an organiser. A client will not place their trust in qualifications but in the demonstrable experience and knowledge of the organiser. Likewise, an organiser cannot convey trust to a potential client by qualifications alone but will need to demonstrate their portfolio of clients and successes. This is where accreditation from professional associations plays its part because clients might be impressed (at least, in part) with an organiser's membership of 'authoritative' bodies.

2.1.3 Qualifications

There is a noticeable shift in the developed industry's recognition of qualifications since universities in the UK started providing degrees in events management. Recruiters are now requiring a relevant qualification when advertising their job vacancies. This in turn requires applicants to meet that requirement. Thus, entrants to the industry need to be more employable and are seeking knowledge of organising events while gaining a recognised qualification. There is a plurality of learning needs here, with entrants in the developed industry requiring both practical skills and theoretical knowledge. Students of events management instinctively recognise these two needs, but Higher Education Institutions (HEIs) are slow to respond. In fact, HEIs have moved from providing practical skills because it is not considered 'academic'. Although work placements in industry during an events course are somewhat fulfilling the need for practical experience alongside an academic qualification, where an HEI does facilitate practical skills, the institution will be serving the needs of students and the industry which employs their graduates.

While junior entrants in the developed industry can reproduce knowledge from current senior organisers, it will not be knowledge gained from the source of formal education because it was not available to senior organisers at the time when they set out to organise events. Because of this, junior entrants might learn poor practices or omissions of certain procedures from their managers.

The same is true in the emerging industry because senior-level organisers will not have had access to formal education, so junior entrants will reproduce knowledge from the way senior-level organisers are doing things. However, the new generation of organisers in the UK who are graduating with qualifications in the art of managing events will produce new knowledge to implement with their future roles and transfer it to their senior organisers, thus importing formal knowledge to the industry:

> *One of them has a degree in event management – one of my event planners in my ops* [operations] *team has a degree in event management from Bournemouth. . . . I have worked with people who have got an event management degree though in various different roles.*
>
> (UK14 – portfolio of royal museum venues)

This organiser presents evidence of knowledge transfer from a team member who brings formal education in event specialism to the team. This is a key factor for positively manipulating the shape of an events industry by educating junior entrants in the specialisms of managing events, who then enter the industry and transfer their education to other practitioner organisers and senior organisers who are organising events with knowledge only. This demonstrates how the developed industry shifted shape from a previous stance of abstaining from formal education and professional qualifications. An emerging and developing industry will organically develop in the same way, but it would be helpful if this were known so that the shape could be consciously steered this way. It would be a quicker transition to providing courses to junior entrants who will import their formal education to current organisers.

The following excerpt is an even clearer example of junior entrants importing their formal education to the UK events industry to influence senior-level organisers:

> *We're quite an old department – except for one person we're all over 50 and that means our education is a mix of formal and operational experience – but formal in hotels rather than events. The reason for that is that when I went into education, as far as I remember there were only two universities that did formal degrees in events – and the whole events education format has just absolutely mushroomed and now there are a huge number of people who go and do events in universities all over the place. One of the challenges that we find is the dynamic between us as older people and the type of organiser-client we are encountering and providing service for. They are half our age and come from a very different background of A-levels straight into a university events degree. It's a very interesting dynamic between a 22 or 23-year-old who has the title 'event manager' and myself who is 58 who has the title 'event manager' and we are two entirely different people. You have to be quite careful with your language when dealing with a client who is half my age in that you can't be demeaning in any way shape or form because they very much take exception. Strangely, you are managing the relationship before you are managing the function.*

(UK15 – landmark event hotel)

This organiser recognises being taught by new entrants who are *organiser-clients* bringing their events business to the hotel. It demonstrates a plurality of learning because the former organiser was learning from educated team members, whereas this organiser is learning from educated clients. Because clients are now educated in events management, there has had to be a re-evaluation of the organiser/client relationship. This is about having to adjust to doing the job in a new way, led by junior entrants to the industry.

So, there is a culture shift in the UK events industry brought about by formally educated entrants interacting with current senior organisers and altering how they had previously learnt to do their job.

Poland can, to a certain extent, reproduce knowledge from current senior organisers, but again it will not be knowledge gained from formal education. More problematic for the Poland events industry is the lack of production of new knowledge because there are no graduates with events specialisms to import to senior-level organisers:

Actually, when I came to you – when we met in Warsaw – this was a good opportunity for me to actually see it from the other side – not from the amateurs' side but from the professional side, which was very good for me. So, now I'm in the business, so to speak, as far as our events are concerned.

(PL4 – organiser of charity events)

This organiser confirms their gaining of knowledge from a professional, which otherwise would not have been available to import to their role as an organiser of events. Even though this organiser was already in the business of organising events, they self-identify as being on the *amateurs' side* and viewing the other *professional side*. If this organiser did not have the opportunity to work alongside an event *professional*, as many other organisers in a developing industry will not either, their role of organising events would not have been seen from the professional side. This reveals the gap to fill with educating organisers working in an emerging and developing events industry.

Another gap in knowledge is that Poland organisers tend not to enter the industry with experience in a related field. Organisers in the UK at least import transferrable knowledge from their experience in related fields, such as hospitality:

I went to Ealing [Ealing College of Higher Education, now the University of West London] *for my HND, HCIM* [Higher National Diploma in Hotel, Catering & Institutional Management]. *I did my placement at the Grand* [Hotel]*, Eastbourne. I was a chef at Country Inns for 18 months, was at the QEII* [conference centre, London] *for two and a half years, a restaurant manager with Hilton for two years, and have now been with P&G* [Payne & Gunther catering] *for 22 years.*

(UK1 – landmark arena venue)

This organiser has a typical entry route for senior-level organisers in the UK, because events management at that time was not a defined discipline but was closely associated with hotels, hospitality, tourism, and leisure management. Recently, however, there has been a refined and focused growth in events education to meet the rising demand for events professionals (Getz and

Page, 2016). Events management is now accepted as a quasi-profession in the developed construct because of the increasing number of graduates from educational programmes as well as holders of accreditations from professional associations (Getz and Page, 2016). When in the turn of time graduates elevate to senior roles, carrying with them their formal education and knowledge in organising events, the developed events industry will organically self-professionalise as an occupation based on advanced complex knowledge (Macdonald, 1999). It is clear, then, that the tension between academic qualifications and experience has diminished in the developed industry.

Macdonald (1999) further makes the point that professions are possible only when knowledge emerges as a sociocultural entity in its own right, which is evident with the UK events industry in recent years. This is why some firms and individuals have become trusted and their word and abilities are considered more significant than qualifications because they can demonstrate knowledge, experience, and successes. Still, having an education in a related field, but not in events, is not a disadvantage in the ability to manage events. The skillset of a hospitality professional closely matches that of an events professional such as flexibility, adaptability, communication, social and human skills, and decision-making (Bowdin *et al.*, 2011). Bowdin *et al.* further identify project management skills as being key to the role, including developing and working in a team and providing leadership, integrating the project plan, presentation, negotiation, and defining client requirements.

Project management is defined by The Project Management Institute as 'the application of knowledge, skills, tools and techniques to project activities to meet project requirements' (PMI, 2013, cited in Haniff and Salama, 2016, p. 12). And the career path of the following organiser demonstrates that project management is closely related to venue management:

> *I didn't study event management. I was working part-time at the college and just got more and more involved in that, really. I was asked to step in on a full-time basis very much as a coordination position and then just worked from there. I had the opportunity to work as the event manager at Alexandra Palace for just under 4 years. I then joined Earl's Court as an event manager. At Earl's Court I slowly progressed through the ranks of management – soft-services management, that kind of thing. Then I was asked to be head of events at Earl's Court during the Olympics until we closed the venue. When Earl's Court closed, there was already a head of events at [name of venue] so a new role was created as deputy director for operations looking after all core services interacting with events, looking after everything from catering, security, traffic and logistics, car-parking, media.*

> (UK10 – landmark exhibition centre)

This organiser can be considered a project manager rather than an event organiser but the transferable skills are identifiable across those two roles anyhow. More pertinent is their knowledge in the sector of managing events and venues.

There is acceptance both in existing event literature and in the events industry that a formal education in a related field such as hospitality or tourism is relatable to managing events. Bourdieu (Grenfell, 2012, p. 103) refers to this as 'institutionalised education' in an attempt to develop a habitus where people know the 'rules of the game'. This dissolves tensions between experience and academic qualifications as an events industry pivots towards professionalisation. However, hotel management is not events management, which raises a concern with what rules of the game of managing events might be unknown if an organiser is not educated in how to play it – if they have studied hotel management, say:

> *No. I don't* [have any formal specialist training or education in managing events]. *No, I don't think* [my team] *do. I'm sure it's not the answer you're wanting to hear. Certainly, the two most senior people in that department have years of experience* [pause] *in events.*
>
> (UK11 – country house hotel)

This organiser accepts that there is a shortfall in their venue because nobody within the events department has specialist training or education in managing events. The interviewee feels conscious of the underqualified personnel in the events department at their hotel. But by recognising the need for academic qualifications alongside experience, this organiser is evidence of the shift in the industry from abstaining from professional qualifications in organising events. This is in fact a glaring admission because events are a third of all businesses in this hotel (the other two-thirds are accommodation and food and beverage). Having stated that the two most senior people in the events department have *years of experience in events* is empirical evidence of trusting individuals, their word, and their abilities more significantly than any qualifications.

However, experience still is not a substitute for having formal education in the specialisms of managing events because the years of experience may not be the best way of doing things. Plus, there will exist the lack of theoretical understanding in factors such as consumer psychology and marketing behaviours, as well as self-recognition of the need for setting event objectives, the value in conducting after-event evaluation procedures, and enacting risk management protocols: every time.

The next interviewed organiser is at another hotel property, which hosts around 300 events each year and considers its hotel and events operations to be *two separate businesses, each with its own entrance*, yet this organiser also demonstrates their lack of specialist education:

Not really. At college and various other times I've done a lot of casual work on events – bar work, plated service, silver service back in the day – all those sorts of things. . . . Just a lot of operational experience. So, no, I don't have any qualifications, but I do have a lot of common sense and practical experience, which is the best thing anyway. You need that, you know. Learning on the job and also seeing a lot of events happen when I was at college – the flow of things and just through experience of doing bigger and bigger events in venues, you learn what works, what doesn't work. A couple of [my team] *do* [have qualifications] *– a couple of them have degrees in event management.*

(UK12 – landmark event hotel)

This is a landmark hotel in central London where catering revenue from events is higher than total rooms' revenue. This is an events-led hotel because room occupancy is low without events. Events are a significant contribution to the profitability and sustainability of this hotel business, yet this senior-level organiser has no specialist training or education in managing events.

It might seem odd that people without an education in organising events can lead an events department which has significant contribution to the business, but this is typical of an events industry at any stage of development. It only demonstrates how the organiser has reached their position because of experience, not qualifications. This is another factual of the industry reinforcing the trust in an organiser's abilities more significantly than any qualifications.

Leaders in events might feel inadequate within their changing socio-professional culture that is now recruiting graduates in events management. However, senior organisers welcome imported specialisms with graduates and there is no evidence of resistance due to change, insecurity, or vanity. This is because senior organisers are embedded in their seniority and feel comfortable and secure in performing their role – again, through recognition of their experience and continued successes and not for qualifications. Senior organisers, then, recognise the value of their self-learning environment and accept they can develop and adapt to the changing culture of their events industry by learning from qualified new entrants.

In the events industry taking shape in Poland there is no access to formal institutionalised education in events management, thus no habitus has yet evolved where people know the 'rules of the game' (Grenfell, 2012, p. 103). In fact, Poland is at the stage where the UK used to be in eschewing any formal education or qualification in events management, driven by a belief that if specialist education is not available, it cannot be needed. It does make sense, then, for the events industry in Poland to place trust in the word and abilities of firms and individuals more significantly than any qualifications which are not yet available, anyhow. The ones who are trusted are those with a portfolio of clients, testimonials, demonstrable experience, and a track record of previous successful events.

This means that organisers in a developing culture must design their own 'rules of the game' and play it their way. This creates an obvious disparity between the game they are playing in the emerging location, and the game the rest of the international events industry may be playing. In other words, there is a different set of rules and a second tier of standards of events within the one global market for international events.

The gap here is caused by the progressing stage of the events industry in the emerging location, which is why, in Poland, experience from any field is valued above specific experience, or even experience from a related field:

Well, I started working for the sports and international sports cooperation department within the local community town hall which used to have the only big stadium in Poland for a long, long time before Euro2012 arrived in Poland and gave the boost to the construction of many other modern arenas. I dealt with lots of cultural events, international events with our partner cities – obviously it was a smaller local scale but still it was the whole logistics and the whole schedule and preparation process of the events. Later on, I went on to a higher-level regional administration level where I was the head of the Euro2012 preparation team for a year. After that year, I went on to Warsaw to the Polish Ministry of Sport responsible for preparing infrastructure for Euro2012. We had the project management infrastructure for highways and mainly for the construction of the new stadiums [sic] for Euro2012. We had realised the project of Euro2012 in Poland was not only about having fun and enjoying 3 weeks of tournament, but it was about giving a boost to the economy, new roads, new stadiums [sic], hotels, huge impact on Polish development and growth which was the main aim, apart from having fun watching football. After that for two years I worked at the National Stadium in Warsaw responsible for event management and deputy head of the sales department. After that I moved to Katowice because a brand new international congress centre was commissioned and they were looking for the operator so I started working for the company that won procurement responsible for that company operating events.

(PL6 – landmark convention centre)

Like the pathway of current UK senior-level organisers, this Poland organiser demonstrates how they entered events without specialist education in managing events but had *started working for the sports and international sports cooperation department within the local community town hall*, which brought them close to events. This exposes another reason why the industry eschews any formal education or academic qualifications in events management: because senior organisers themselves have succeeded to their position without the need to be educated or qualified in the discipline. They are wont, therefore, to reject the need for others to be formally educated or qualified.

The above organiser found themselves to be *the head of the Euro2012 preparation team for a year*, which is a significant leadership role in the field of events management. Then, they talk about the *project management infrastructure for highways and mainly for the construction of the new stadiums* [*sic*] *for Euro2012*. Analysing the career trajectory of this organiser shows that they are less an event organiser and more an infrastructure project manager for events. There is nothing wrong with that, because events do require infrastructure project specialisms in their art world. But the point is that this organiser does not have the education in events management to make those infrastructure decisions that will impact an event.

In the above excerpt, the organiser acknowledges the economic and infrastructure benefits for Poland from hosting the UEFA Euro2012 football championship due to which they gained learning in project management. This highlights the point that the production and reproduction of knowledge can get going where cultures provide opportunity for policy transfer to learn from each other. However, it also indicates that the value of events education as a discipline is not yet recognised in Poland.

All is not lost, though. The events industry in Poland is still taking shape, and the beginnings of courses in events management are now being offered at both the University of Warsaw and the University of Commerce:

[Talking about universities] *A new thing which has just been created and we have no idea whether it is any good.*

(PL4 – organiser of charity events)

Poland getting going with a culture of knowledge reproduction and production (Bourdieu, 1973, cited in Harker, Mahar and Wilkes, 1990) is a sign of the opportunity to begin reducing the knowledge gap in the emerging industry with other players in the global market for international events whose socioprofessional habitus has already taken shape and is producing graduates to import specialisms into their events industry.

Still, the above interviewee is a current organiser in a capital city, and who said they have *no idea whether* [the events course] *is any good*. Clearly, they have heard of the new events course but have not taken any effort or interest to find out about it or if it *is any good*. This indicates that, first, the organiser is not fully immersed in the socio-professional construct of their profession to find out more about the new course. Second, they feel no need to take the new course that might now teach them how to do the job. Third, they have no recognition of their need to create for themselves a self-learning environment. This Poland organiser, then, is eschewing any formal education or qualification in events management.

Here is another indication that the emerging construct is already shaping a fragmented autonomous approach to delivering events:

This industry over the past years has not been integrated because the number of the big arenas and congress centres like mine has increased over the last 5 years. Before that there was a shortage of them. Within those past 5 years the industry has not been able to integrate yet to lobby on government level for new laws to support the situation. But I think it was last year the organisers of concerts, gigs, entertainment shows and big festivals such as the Warsaw Orange Festival have come together and they set up an organisation in order to exchange knowledge and in order to operate or to work out some common policies towards venues, clients and government.

(PL6 – landmark convention centre)

This organiser confirms what the previous organiser indicated: that the industry *has not been integrated* and *the industry has not been able to integrate yet.* But it does show that the emerging industry is moving towards collaboration and there is a sociocultural shift towards a community of practice happening, with all the benefits this brings through socio-professional networking. In this case, it was the arrival of the COVID-19 pandemic which caused a crisis in the events industry and triggered the need for organisers to collaborate because of the sudden and urgent reliance on cooperation from others (Becker, 2008):

What we have seen because of the COVID-19 situation . . . pushed us to exchange knowledge on our situation and how we can influence decisions the government is going to take to support our industry. It was probably the first time that I had the chance to come together to talk with the main arenas and congress centres because we wanted to support and give some input to people representing different entertainment and sport associations to talk on our behalf to our government to work out best solutions in these difficult times. It was the first time we were exchanging knowledge and we working on documents to propose to our government and was also the time we realised how important this type of cooperation was. . . . I think this is something that will lead in the future to set up a body that would assemble representatives of venues which would be helpful.

(PL6 – landmark convention centre)

The COVID-19 pandemic forcibly advanced the events industry in Poland into developing a culture of knowledge sharing, although there are indications this was happening anyway. The organiser above frames the dramatic changes caused by the arrival of the pandemic as the catalyst for collaboration and that they found it both necessary and helpful to network with other venues which would have been treated as hostile competitors prior: *It was probably the first time that I had the chance to come together to talk with the main arenas and congress centres*. This reaffirms that competition is healthy and should not be held as a barrier to collaboration within the community of practice. Smaho (2012) states that knowledge sharing is a consequence of and influenced by dramatic changes in economic and social conditions, as well as changing consumer demands and requirements, which can certainly be applied to Poland since the collapse of communism in 1989 and then its integration to the EU in 2004 but also by the sudden impacts of COVID-19.

The pandemic may have been the driver to speed collaboration in the Poland events industry, but organisers in an emergent location will need to maintain that urgent momentum to move rapidly in catching up with the advanced state of developed locations in the global events industry. Frankel (1979, p. 9) calls this 'orientation towards participation in the international system', which is the core objective of the potential for an emerging events industry to orientate towards participating in the global marketplace for international events.

However, it does not always work like that in real terms – that it is an 'orientation towards participation in the international system'. That is because emerging locations tend to happen suddenly and quickly, which is one of the factors that cause the gaps within the one marketplace. Qatar hosting the 2022 World Cup, or when Poland and Ukraine hosted the EURO2012 football championship – these locations found themselves thrust into the market for international events rather than oriented towards participation in the international system. There is not always, then, the luxury of orientation before events in an emerging location are already taking place – it is this which defines it as being an emerging location in most cases. Rather, it is orientation towards participation in the international system *after* a location emerges into the market for international events: this is wherein the opportunity for orientation arises.

For this, organisers in the emerging location need to recognise their need for collaboration with foreign sources in order to take advantage of differences in expertise and to access markets around the world (Argote *et al.*, 2000). Bourdieu (Grenfell, 2012) explains that this can happen in terms of competition for capital advantage in social spaces or 'fields' such as economic capital, cultural capital, knowledge capital, and social capital. With both willingness and the recognition of the need, the emerging industry can adopt legislations from the developed industry because the younger events industry is at the point of opportunity to advance to professionalisation.

Domański (2003) highlights that Poland is pushing closer to the standards and structures of other EU countries, because progress in organisation and management is driven by greater competitiveness, so the motivation is already here. There are, however, embedded societal and cultural characteristics in Poland which are barriers to the rapid development of the Poland events industry through collaboration. Rojek (2013, p. 10) mentions 'social ordering' where there is a process of conditioning social behaviour through formal (education; law) and informal (understandings; can-do attitude) processes. This is observed in Poland with advice being rejected because 'you might do it that way in London, but it won't work here', or 'that's not the way we do things here'.

This factual appears at odds that there is trust in the word and abilities of an organiser. However, this is a sociocultural difference rather than about the significance of qualifications: the two are very different. Beckford (2017) states that expressions such as 'We've always done it like that' is an indication of systems and procedures becoming fixed or frozen and the pressure for change meeting high resistance. Although this type of resistance is not limited to Poland, it is a sociocultural attitude influenced by Poland's communist past and is reflected in research undertaken by Dobosz-Bourne (2004) into the different processes of *Opel Polska* and *Vauxhall Luton*. It showed that the same set of practices can become the norm and a source of power in one location or be rejected as a negative practice in another location. Even so, Lipton *et al.* (1990) found that there is an intense desire in Poland to rejoin the economies of Western Europe for the obvious achievements as well as a race to distance the failures of communism, making it the first country to enact fundamental market reform under a non-communist government.

In Poland, there is a cultural distrust of foreign intervention: cultural pride to reject collaboration with an outsider's way of doing things. Giddens (2012) concurs mistrust as being an actively negative attitude towards the claims to expertise, and Mishler and Rose (1997) identify distrust as the legacy in post-communist Europe as a cause of resistance to outside interference, which seems to be the case with the rejection of professional expertise.

This aside, the frankness of Poland organisers is helpful, which is underpinned in this response to the question, 'how do you know you have done a good job':

> *You are the event organiser so you know when an event has worked. You know immediately after this. We have feedback from clients. Usually they are pleased, I must say, with our work. But the Polish companies say, 'Thank you, goodbye'.*

(PL5 – landmark historic venue)

This interviewee, and others, demonstrates a pattern for organisers considering themselves to be the best judge of doing a good job. This is without conducting structured after-event evaluation procedures which would inform an organiser from stakeholder feedback and performance measures.

Organisers place trust in themselves because of their own experience and abilities, rather than placing trust in the learning they would achieve if they were conducting after-event procedures, feedback, and measures against objectives. It is evident that the above organiser does not recognise, for instance, the value of conducting after-event evaluation; otherwise they would be doing so. In this case, they would not rest on the feelings they get to inform if they have done a good job.

None of the interviewed Poland organisers conduct after-event evaluation but this is not a socio-professional or cultural difference from the UK because most of the UK organisers do not do so either. Some of them do, however, so there is more after-event evaluation happening in the UK than in Poland. But this is not because of socio-professional or cultural differences, or because of learning from formal education. Those UK organisers who are performing after-event evaluation do so for reporting purposes and because UK organisers have experience of conducting after-event procedures and so it becomes embedded in their practice.

2.1.4 After-Event Evaluation

In the previous excerpt, the organiser said, *We have feedback from clients. . . . But the Polish companies say, Thank you, goodbye.* This reveals Western clients from the developed industry expect to give their feedback but Polish clients in the developing industry just walk away at the end of an event. This is because Western clients are routinely being asked for feedback and are used to giving it, so they do so when in Poland as well.

Poland organisers should be learning from the differences of their clients from a developed marketplace. But even not obtaining feedback from their domestic Poland clients is a concern because the organiser has no measure of their performance from the domestic marketplace to determine whether they are doing a good job.

This reveals complacency with obtaining feedback because this organiser happens to receive it from Western clients only because it is offered, not by routinely conducting procedures with all clients to obtain it. This is where the value of after-event procedures is not being recognised because understanding the value of doing so means it cannot be overlooked and must be carried out as a routine procedure at every event, every time. It is what is referred to in Berners (2017, pp. 149–150; 2019, p. 188) as performing the job of an event organiser in its entirety whereby pre-event planning is 95 percent of the job, the onsite logistics is 3 percent, and the after-event procedures are 2 percent – but if the after-event procedures are not performed, the job is only 98 percent complete.

Entrants to the UK industry who have an education in events management (because of its developed shape) are being taught to recognise the value of after-event evaluation and feedback. So, there is change happening in this socio-professional practice of managing events. Poland demonstrates this is not yet happening in the shaping stage. This is confirmed by another organiser who measures their performance from a reactive and subjective viewpoint by interpreting positive but informal feedback as affirmation that they are doing a good job:

> *If our clients are pleased with what we do, if they are really pleased with*
> *that, that means we think we have done a very good job.*
>
> (PL4 – organiser of charity events)

For this organiser, to *think* they have done a good job is not the same as knowing it to be factual. If they were conducting post-event evaluation and feedback, they would not use the word *think* but would have full confidence in the fact and would have wanted this to come across in a research interview.

There is further ambiguity revealed in this short excerpt with their use of the word *if* in *if our clients are pleased with what we do*, which indicates that their clients are not always pleased with what they do or that it is not known in every circumstance at every event. Underpinning this interpretation is the certain knowledge that this organiser does not conduct structured evaluation procedures.

There is further evidence of organisers trusting in the 'feeling' of doing a good job, which appears important to organisers in the developing industry and is again based on informal feedback with no structured after-event evaluation processes taking place:

> [Sigh] *How I measure it? I think basing feedback is from the client and*
> *internally – both ways. I've done a good job because I haven't had com-*
> *plaints. Everything was ready on time. Um, the client wants to come back*
> *next time. The area that I was responsible for worked effectively and this*
> *is how I feel.*
>
> (PL6 – landmark convention centre)

The sigh at the outset of this excerpt could be interpreted as a delay tactic, whereas if structured post-event evaluation were being conducted by this organiser, they would perhaps respond without hesitation. This interpretation might further be underscored by the interviewee repeating part of the question, *How I measure it?*

This organiser assumes that they have done a good job if no complaints were received, which is counter-intuitive to having done a good job because there would not be any complaints, anyhow. The purpose of doing a good job, after all, is to receive no complaints, to pre-empt potential complaints, and to take preventive and corrective measures before it is too late: before complaints become complaints (Berners and Martin, 2022).

Receiving complaints as a measure of performance is not the same as conducting the after-event procedures of feedback and evaluation to confirm to the organiser that they have done a good job. For one thing, not everybody will complain: there are many reasons why people do not complain. Also, people might not complain to the event organiser (or even be aware of who *is* the organiser) but might raise their concerns to somebody else within the organisation or outside, so the organiser would not get to hear about it. Furthermore, the purpose of planned structured feedback is to penetrate all stakeholder groups – so the organiser might not receive a complaint from a client, say, and thus feels a good job has been done, but this does not account for consumers, sponsors, suppliers, and staff.

But the overriding value of post-event evaluation and feedback is to learn whether the job was done well because the event has met its objectives and to learn what could be improved – it is not about waiting to learn from complaints.

These excerpts from the interviews confirm that organisers in the emerging industry are not planning to meet objectives such as client satisfaction, delivering to consumer expectations, or achieving repeat business. If a client or consumers were not happy with an event, or did not wish to return, there is a real risk that the organiser will not know about it.

One problem is that audit practices such as evaluation can be seen as mundane, but they create the larger picture which takes on contours and identifies patterns (Strathern, 2000). Beckford (2017) highlights that learning results from a systematic cycle of planning, experimentation, reflection, and consolidation, which supports the structured approaches that UK organisers describe in the developed industry:

I'll answer it in two ways. From a commercial point of view has it delivered to its commercial objectives. Yes or no. The second way, we send out a survey after each show to get feedback from our organisers. If they're not happy they'll tell you. The other thing is I spend time with my clients and we'll have non-commercial meetings – we'll leave the contracts and money and have a qualitative meeting and say right let's forget the money side of it, where are you with your event? What do you want to do? How was the last event? If it didn't go well what can we do to help you to get it better – is it something we can do; is it a connection we can help you with; is it a bit of knowledge; is it someone in the industry who might be able to help?

(UK4 – landmark convention centre)

This organiser conducts non-commercial meetings with clients before an event which demonstrates learning in and between projects to develop competence in ordinary activities (Clegg, Skyttermoen and Vaagaasar, 2021). This organiser is clear about the importance of meeting commercial objectives but separates the non-commercial objectives to allow for the understanding of how an event can be successful in other key areas. This approach is in line with the opinion by Getz (2018) that event managers need to implement continuous evaluation to become learning organisations, achieve their goals, and meet the standards expected of a profession that provides services to society.

Most of the interviewed UK organisers undertake after-event evaluation in some form, compared to none of the Poland organisers, which confirms that in the emerging market there is wider non-understanding in the value of evaluation. It also reveals that there is a failure in recognising 'evaluation' as industry best practice which exposes this gap is due to the stage of evolution of an events industry.

Collaboration within a developed socio-professional structure (Becker, 1984) would redress this, as is already apparent in the UK. There is no evidence to show that this gap is caused by cultural or social differences, it is only about organisers learning the value of conducting after-event evaluation, in which the UK respondents are more aware is a measurement of performance:

> *Obviously there are the financial ones* [KPIs]. *And then what we do pore over are the Net Promoter Scores – we just refer to them as NPS. It is very much a tool used by a number of media-type people. If you can imagine that you're asked the question after an event, 'how would you rate the event and would you come back' if you score (out of 10) between 9 or 10 that's a plus. If you score less than 5 it's a minus. Between 5 and 8 is considered to be neutral. We have a vast database and we send this question out to hundreds, if not thousands, of people saying 'would you recommend to come back next year'. Then we collate what all the scores are. You get a NPS whereby you've got all the 9s and 10s and you deduct all the 5s or less, and you'll come up with a plus or minus. And that's the key thing we use as a KPI going forward.*

(UK3 – landmark sports venue)

This organiser puts reliance on Net Promoter Scores (NPS) as a tool to evaluate and measure consumer feedback responses. This approach concurs with Goldblatt (2011) asserting that evaluation is connected to setting Key Performance Indicators (KPIs) for an event. Badawy *et al.* (2016) explain that KPIs are mostly quantitative information important for planning and controlling, creating transparency, and supporting management decision-makers. Measuring performance through KPIs and evaluation leans towards an audit culture which has shifted from finance and accounting to all kinds of reckonings, evaluations, measurements, and accountabilities (Strathern, 2000) and has social consequences in locking up time, personnel, and resources. This is a barrier for organisers who want to conduct post-event procedures because of the anti-climax effect once an event has gone, plus the drive to move on to the next event for commercial reasons.

Strathern (2000) points out that an audit culture creates problems with checking and evaluating, such as the requirement of trust in the measures used and the sources of information. In this respect, NPS can be negatively impacted even by the weather because if it is raining on the day of an event, the consumer experience will be compromised and the NPS would be lower:

What we've learned for example is that Net Promoter Scores are governed by the weather. In an event that's predominantly an outside event, however good your food and beverage service might me, however good the actual event is, if it's the British summertime and it is pouring down, then your NPS drops. So, when we review our annual NPS we actually have above it 'good weather, bad weather'. But we are always looking for high plus scores. . . . You will get companies – some budget airlines for example – they have got an NPS score that is about minus 70 or whatever, so we're always looking for a high plus score.

(UK3 – landmark sports venue)

The audit culture in the advanced UK events industry is especially pronounced, whereas in the Poland socio-professional culture of events, which is still forming, it is largely non-interventionist. This means that an event happening in the UK faces stringent audit culture processes which add time, logistics, and money to comply with, yet events in Poland might not require such resources to be applied to be compliant.

2.1.5 Event Objectives

Allen *et al.* (2010) consider evaluation as a critical step in the planning process of an event where objectives set at the outset become the benchmarks to determine its final outcomes and success in delivering a service to clients and consumers. Not all UK respondents recognise this because some evaluate their events without setting or using objectives to measure success:

Not measurable objectives, no. Every event has a detailed schedule – if this is met, we know we've done a good job.

(UK1 – landmark arena venue)

This organiser is confused with what an objective is because an objective must be measurable by definition. The *detailed schedule* to which this organiser refers is otherwise known as an 'Event Management Plan' (EMP), 'function sheet', or 'running order' but is not produced for measuring objectives. An EMP is a logistics document used for communicating the schedule of an event to interested parties and, yes, for that reason it does help in ensuring the objectives of an event do get met, but that is not its purpose. Besides, meeting the content of a schedule might mean the event ran smoothly but does not necessarily mean a good job was done because this will only be determined from the after-event evaluation and feedback audit procedures.

Another organiser stumbled over the question of setting objectives:

No; no formal objectives. It sounds weird but I suppose we just know what a good event looks and feels like, and what needs to be done, and what we need to achieve but nothing formal, so nothing written down in an event plan. . . . No. It's a good question.

(UK2 – organiser at an institution)

This response again introduces confusion and ambiguity, yet the question, 'Do you set event objectives for every event?' is clear and simply worded. In this case, a follow-up question seeks clarity by asking, 'If you are not setting formal objectives, how do you know the event was a success?':

Hmm. [Long hesitation] *That's interesting, isn't it? And it's different for different events.* [Long pause] *That's, that's really interesting.*

(UK2 – organiser at an institution)

The statement, *It's different for different events*, is acknowledgement that objectives need to be set for each event as a routine procedure, but this organiser does not do this. This organiser then went on to list a range of potential audits for the different events they organise, including:

Good levels of attendance; no technical problems; people stayed for the drinks reception afterwards; do people look happy after the event; everybody is mentioned in the speech; the right people are in the front row.

The organiser could easily have answered 'No, I don't set objectives', but their floundering, hesitation, and reaching for a list of possible factors, which could be used as measures, indicate their recognition of their failure in not setting objectives.

Poland is not dissimilar to the UK, where organisers know it can be important to set objectives, but fail to do so:

[Hesitates] *Well, yes but not always on paper. In Poland the majority happens during conversation.* [We] *set objectives verbally. Objectives are less visible than they should be, not necessarily put in writing. That's how it works, if you know what I mean. I don't believe anybody in Poland does such a thing. Honestly, in Poland we don't do it, we don't measure it because there is never time after an event.*

(PL7 – organiser of events)

This leads again to the conclusion that not conducting strategic management practices is not an omission caused by cultural or educational differences between the emerged and the emerging industries. It is purely non-understanding of what an objective is and what it is for.

This organiser states they do not *believe anybody in Poland does such a thing*, and *Honestly, in Poland we don't do it*, suggesting that just because this organiser does not set objectives, they perceive nobody else does. This confirms how organisers assume that the way they do things is *the* way of doing things, but it might not be the best or safest way.

This organiser also says, *we don't measure it because there is never time after an event*, which is the truth reveal. Organisers in both the developed and developing industries blame lack of time after an event for not conducting post-event procedures (see also Section 4.6, 'The Self-Learning Environment'). Lack of time is the reason proffered by organisers to explain why evaluation is the neglected part of the event cycle (Quick, 2020), so this research underpins what is factual in the industry. However, there is always time for strategic management practices – it is part of the job.

Lack of time is not a legitimate reason because although evaluation activities take place after an event, they are planned before an event – in the planning stage. It does not require time from the organiser after an event to plan after-event activities nor does it take time to act on them – the consumer does that: it is the customer who sits down to complete an evaluation questionnaire. It is pointed out in Berners and Martin (2022, p. 53) that 'time does not prevent feedback and evaluation – poor planning does that'.

The interviews with organisers reflect a *laissez-faire* culture exists within the events industry in both the UK and Poland. This reveals itself in strategic events management but is also visible in both the Event Tourism and Risk Awareness sections of this book. This is more prominent in Poland than the UK:

> *Of course* [I set objectives] *when I am the organiser because sometimes the* [venue name] *is not only the venue it's the organiser or co-organiser of the events. But when the* [venue name] *is a venue, it depends on the client, that's their job to set the objectives.*
>
> (PL5 – landmark historic venue)

This organiser sets objectives when organising events in their venue and, by using the term *of course* at the outset indicates they attach value to setting objectives and understands it is an essential part of the event management procedures. Even so, this organiser is displaying the *laissez-faire* characteristics by foregoing the need to set objectives in situations where clients organise an event in the venue (known as 'dry hire'). This reinforces how organisers in both the developed and developing industries do go some way to setting objectives and conducting evaluation but do not fully recognise the value of doing so, otherwise these activities would not be foregone in any instance or for any event. Neither do organisers appreciate the relationship between objectives, KPIs, and evaluation as metrics of their performance (Jeston, 2018) to inform whether they have done a good job.

2.2 Socio-Professional Culture

Berners (2017, p. 3) categorises the complex socio-professional culture of the event industry in developed markets such as the UK as 'an industry in itself'. This is because the structure of the event industry in the UK is multi-layered with event management agencies, production companies, freelancers, and the many services which support a developed event industry, from marquee suppliers to power supply companies, temporary trackways, perimeter fencing, and crowd barriers. In essence, the size and structure of the socio-professional network which has evolved to support an events industry are what determine the shape of that industry and where it is positioned on the journey from emerging to developed.

The events industry in a developing market is a different shape: it has less need to develop support from a socio-professional network and so has fewer layers which make it a simpler and 'noticeably flat' (Berners, 2017, p. 5) shape. This is certainly what is observed in Poland because it can be difficult to source creative resources, there is limited choice of support services such as technical companies, and there is no existence of professional associations to network with suppliers. In fact, in Poland there are few 'event organisers', but events are being organised by people and companies involved in marketing, public relations, and media.

There are voids in event specialism in both Poland and the UK which get filled by organisers with limited or no specialist experience. Organisers without specialism can be devoid of the event management skills required of the event life cycle (Holmes and Ali-Knight, 2017) which includes planning events, setting event objectives, marketing events, optimisation and growth of events, safety protocols at events, and after-event evaluation.

2.2.1 Void of Organisers

Although it is rapidly evolving in Poland as the events industry takes shape, and there are now event agencies which are shaping structured levels to the Poland events industry, in an emerging events industry there is a void of specialist event professionals. This causes gaps in the specialism of organising events that get filled by non-specialist personnel working in related media-centric professions offering themselves as bona fide event organisers:

> *I worked for 15 years in the mobile telecommunications sector. I did some*
> *marketing there and that's why I'm here at* [venue name] *to somehow*

support marketing and to organise events. But most of my career was in a different place and different sector. I do not have any theoretical background but rather I could say practical. Please bear in mind this is not a commercial company. The municipality makes all the decisions here because we have to ask if we can host the events in this [venue name] so most of the decisions are on the side of the municipality.

(PL2 – landmark venue)

This organiser confirms that in the developing industry there is deficiency in the specialist knowledge of managing events. That is why this organiser was awarded their position – because of their marketing background, which was perceived to be related and therefore sufficient to lead events at a capital city landmark venue.

Venues such as this attract big events, international clients, significant budgets, and sometimes complex logistics. Importantly, this type of venue has a public profile and needs its reputation to be protected. In an emerging industry it would be difficult for this organiser's employer to find a candidate with specialist event or venue management experience, anyhow. Best to go, then, for a near-related career and hope there is cross-skilling.

Their comment that *most of the decisions are on the side of the municipality* must limit the ability and authority of this organiser as a venue manager. If they did, however, have specialist experience in events and venue management to give them confidence and authority, they might feel empowered to take decisions in their capacity as the venue manager.

From this perspective – the perspective of organisers – there is similar cultural structure in both the developed and developing industries, as the following two excerpts from a UK organiser and then a Poland organiser describe:

It was a little bit accidental. I was actually studying a double degree in marketing and politics. I was working in a hotel whilst at university and I got offered the position of being the events manager at the hotel and I really liked it. That's basically how I got into that. And then I volunteered in India for a little while with underprivileged children. I was told I had no London experience. . . . [T]hey wanted London experience in particular. So I started off at the very bottom at [hotel name] working on reactive sales and planning of events as well. Then I went into proactive sales, my remit currently is basically ensuring I have event management agencies and corporates coming into our hotel. My remit is the globe, US market, Asian market, Europe market, as well as the UK market knowing about [venue name].

(UK13 – sales account director)

Many, many years ago when the idea of the [event] came to my mind, I had no way out. I had to start organising events. So, it came naturally. I didn't study anything which involves organising events, so it came just naturally from the bottom of my heart and my head.

(PL3 – organiser of charity events)

Both these organisers demonstrate their similar accidental pathway into events management because the events socio-professional status in both countries – and at different stages of development – makes no demand for relevant academic qualifications, training, experience, or specialisms. This provides a snapshot of an events industry, wherever it is, filled with professionals who have learnt to do the job but may not be doing it right or safely. In such cases, the organisers shift from 'dedicated and resourceful amateurs' to 'trained and skilled professionals' (Bowdin *et al.*, 2011, p. xxviii) because of the experience they are gaining and their social practices become characterised by the presence of rules to dictate good practice (Bourdieu, no date, cited in Grenfell, 2012) which is explained here:

We gained experience whilst organising things because we always had to deal with these problems and we always had to see to it that everything was tip-top. There would be no problems during the event. One of the events that we organise is really huge therefore we had to see to everything. And have professionals deal with the problems, we didn't do it on our own. We had firefighters, security, medical people, special guards from the [venue name].

(PL4 – organiser of charity events)

This organiser, like most, has evolved their own mini-culture for organising events which might not be the right way, or the safe way, or the optimum way. This organiser relies on what they label *professionals* collaborating to ensure a successful event, so there is evidence of the use of a wider art world supporting their mini-culture. Becker (2008) recognises that all human activity involves the joint contributions of a number of people and suggests that pattern of collective activity is a sociological approach. This organiser does therefore have a sociological approach, albeit unintended.

Reliance on other people is usual in events socio-professional culture because there is an acceptance that events are run by people who refer to themselves as organisers but without attaching any professional merits such as specialist education or experience to qualify them as such. In an emerging industry this is because organisations are reacting to newly emerging free trade, bringing competition from the rest of the world (Lipton *et al.*, 1990). For example, Domański (2003) recognises Poland firms must adapt to unprecedented competition in a sudden environment of catching up to new demand outpacing the labour resource of organisers who are experienced or educated in managing events.

As an events industry begins to take shape, a culture develops which impedes rapid catch up with developed industries in the global market for international events. This is because there is no access to formal education in events specialisms in a location with a young events industry. Until the industry matures to a point where providers of education respond to a demand for courses in events management, organisers will be recruited into the industry for reasons other than industry-related qualifications – usually it is by chance. Nor are they being offered event jobs because of their track history in organising events – they do not have one:

> *Huh, I worked for 15 years in mobile telecommunications sector. I did some marketing there and that's why I'm here at* [venue] *to somehow support marketing and to organise events. But most of my career was in a different place and a different sector.*

(PL2 – landmark venue)

It is fortunate, then, that such organisers are supported by the wider socio-professional habitus, which includes other industries that can dampen their deficiencies of knowledge and expertise. They can rely on the expertise of other skilled operatives such as specialist caterers, venue managers, risk and safety experts, trained security, technical specialists, and agency casual staff such as waiters. It is a structure which works to cover the gaps within the industry which should be led by event organisers, but in many cases is led by other people with specialist other skills.

This will not be happening to the same extent in a location such as Poland, where the industry has not yet evolved that wider socio-professional structure. In Poland, the industry is developing a culture that is highly autonomous of managerial control and largely immune to formal management regimes (Thiel, 2012, p. 4), because there is no community of practice to cohere or align organisers and narrow the gaps between them.

There is a further widening gap between the UK and Poland culture of events management, due to the correlation between educational investment and income (Domanski, 1999). Domanski asserts that education and job complexity are the indications of social role and status when converting investment in oneself into reward. This is a point that is explored by Kuhlmann (2013) in locations of fundamentally changed concepts of welfare state and the governance of professions. The social and professional approaches to a career in events industry has changed since then, with events in the UK now seen as an industry in its own right (Getz and Page, 2016), which is why this gap is widening.

In Poland there still are very limited opportunities to invest in one's own education as an event organiser (or opportunities to employ educated event organisers for that matter), so qualifications are being sourced from other disciplines. This is indeed the case with the following organiser who is a *historian and an educator* and is now the director of education and events at a landmark historic capital city-centre venue:

> *I am a historian and I was an educator in* [venue name] *which has its educational department. It started with art history students, etc. and we are doing lessons for students attending these lessons. Because of the head of this branch of the education department – she is also the head of the events department – she knows me and when there was a need because the former events director was retiring, she asked me to replace him and jump in.*
>
> (PL5 – landmark historic venue)

This organiser describes a natural and organic progression within the internal culture of their venue, *when there was a need because the former events director was retiring, she asked me to replace him.* Even though this was an appointment of convenience and organic progression, the process neglected the potential to recruit somebody with event specialisms. This organiser freely accepts they were teaching *art history students, etc.*, which does not make them an organiser of events. But the woman who was *the head of this branch of the education department* would not recognise this or know the need for the job to require events specialisms – she just happened to be *also the head of the events department* and there was an opening which this 'organiser' was there to fill. It is unlikely in Poland that an events specialist would be found for this position, anyhow, so why bother looking.

So, the emerging industry eschews any formal education or academic qualification in events management. But until Poland develops their events industry to be a socio-professional culture that provides investment in oneself through education, it will not begin to narrow the gap between itself, the UK, and other developed locations already operating in the marketplace for international events.

It is not only the impact on organisers themselves which is affected by the socio-professional culture being created. The way an events industry takes shape will have impacts in other areas for organisers managing events such as how successfully they penetrate foreign markets to attract event tourists and how well they meet the service and quality expectations of international clients and consumers. It is documented by de Kadt and Williams (1976) that old perspectives linger in underdeveloped countries which has implications in modern society where knowledge-based occupations are integral to modern social formation (Macdonald, 1999). This is a consideration for the development of the events industry in Poland, particularly with its hangover legacy from communism.

2.3 Summary and Conclusions – Event Culture

Exploring the culture of the events industry in both the UK and Poland has identified the characteristics of a developed industry and one that is emerging. Although the two locations differ, in that the UK is a long-time democratic society versus the post-communist nation of Poland, this book is not concerned with such wider cultural complexities, including those of gender, religion, or command cultures. This book therefore limits its focus to the culture of the socio-professional construct of the events industry in both target locations.

The evolved community of practice in the UK events industry has shaped a socio-professional construct to support organisers through networking and activities, professional associations, and accreditation bodies. This is important in professionalising the industry because it provides opportunities for organisers to share best practice and gain knowledge from other professionals within the same industry.

Of particular importance for an events industry construct is to shape a culture of learning, and learning from each other. Organisers should be led to recognise their need for active self-learning in an environment where they learn from sources outside the limitations of their own experience.

As an events industry takes shape, it pivots from eschewing any formal education or academic qualifications in events management. This is because in the developing stages, there is no access to specialist courses which fosters the belief that if qualifications in events management are not available, they are not needed to do the job of organising events. The consequence of this is that experience is valued above qualifications and trust is awarded to the word and abilities of some firms and individuals because of their experience, not for any qualifications.

Trust, then, is based upon an organiser's client portfolio, reputation, experience of events in the genre, testimonials, and track record of delivering successful events. After all, qualifications alone do not convey trust in an event organiser.

Tensions between academic qualifications and experience arise where experienced organisers, who feel they know how to do the job, devalue qualifications and reject qualified entrants who have academic qualifications but lack experience. The industry itself, then, is not leading organisers to be qualified which is why actors in the industry place significance on experience. Once courses become available, entrants to the industry seek academic qualification to make themselves more employable. Recruiters respond by requiring qualified candidates to apply for advertised job roles. Still, this shift to accepting qualifications is led by individuals rather than the industry.

In both the UK and Poland there is a lack of formal education because although the UK is further advanced with its provision of events courses at

HEIs (Higher Education Institutions (universities)), it is still too recent to show much impact at senior levels. Formal education is impacting entrants at junior level, but current senior organisers did not have access to formal education in events management when they entered the industry. However, there is evidence of entrants importing their formal education to the industry which is reshaping the behaviours and practices of senior-level organisers.

Other than learning from graduates, current senior-level organisers in the UK rely on their formal education in related fields such as hospitality, catering, and hotel management. This means events in the UK developed construct are being organised by people who already possess transferrable learning and competences to draw upon, such as communication, cooperation, people skills, taking care of others, working collaboratively, teamwork, and verbal reasoning (Schulz, 2008; Cserháti and Szabó, 2014; Pielichaty *et al.*, 2017; Clegg, Skyttermoen and Vaagaasar, 2021). A gap appears here because organisers in Poland are not entering the events industry from related fields. In the Poland emerging construct, organisers of events most likely do not possess those attributes because they have landed in events management from unrelated professions.

This is not to suggest that there is disparity in professionalism between organisers in the UK and Poland. Rather that although professionalism may be present in the social construct of both industries, in a developing industry it is professionalism gained from an irrelevant profession. Usually, the 'organiser', as they become, has arrived at events management by chance or convenience. However, this gap does mean that organisers in an emerging industry are less attuned to meeting the needs of the global market for international events, than an organiser in a developed construct who brings relevant transferrable competencies.

Without experience in related fields, and the absence of the socio-professional culture from which UK organisers benefit, organisers in an emerging industry evolve their learning of how to do the job in an environment which is both informal and autonomous. This is characteristic with Bourdieu's habitus concept (Grenfell, 2012) where social practices occur as regularities without the presence of rules to dictate practice. Although Foucault (Carrabine, 2007) recognises that strange practices are governed by structural codes of knowledge, observation in the Poland field has shown learning has developed through the way of doing things which does not make it the best way of doing things, or the safest way of doing things.

Although Poland organisers can learn from sources other than formal education or years of background in events – such as from their foreign clients, international consumers, and other organisers' events – there are three barriers to do with sociological behavioural traits that exist because of the legacy of communism. The first is how Poland society needed to turn their hand to whatever needed to be done, which Rojek (2013, p. 10) recognises as informal 'social ordering'. The legacy impact is that it is embedded in Poland society that they can do anything, that everything they do is the right way, and there is no better way of doing it.

The second barrier to embracing outside expertise is the mistrust in one's neighbours. Mishler and Rose (1997) identify distrust as the legacy in post-communist Europe, too. At this level of embedment in society, to ask for trust in the expertise from a foreign source is incomprehensible.

The third barrier is the slowness of societal behaviour change, which continues to make it difficult for Poland organisers to accept foreign expertise. This slowness is because the long reach of communism across the generations repeats communistic cultural patterns and societal behaviours. Although Poland is a post-communist society since 1989, the grandparents who lived under the communist regime taught today's parents, who teach today's children, who will be teaching tomorrow's children. The impact is that changing the informal authority of societal norms is incredibly slow and resistant to change. However, change from the formal authority of education would enact quicker change and with less resistance.

These are cultural barriers in shaping the events industry in Poland to align with the global marketplace for international events and meet the expectations of event tourist clients and consumers. This is problematic because Poland is already a key player in the global events marketplace, having hosted significant international events such as Euro2012 and COP24.

To narrow the gaps between an emerging industry and one which is developed will require a socio-professional culture change across the construct of the global market for international events. This could be accomplished with the creation of an international accreditation association to lead emerging events industries in shaping their socio-professional construct to the same pattern as those which are already established. This would both lead organisers and provide them with the understanding of their need to engage with support structures to facilitate networking, knowledge sharing, formal education, and best practice across the one global marketplace for international events.

In fact, the importance of a self-learning environment, whereby organisers develop a culture of learning from each other, is a finding that is of key importance in the shaping of an emerging industry, which has yet to establish a socio-professional construct. However, self-learning is a trait which current organisers adopt naturally and organically or else fail to recognise.

Currently, then, it will go against the grain for the industry to actively lead organisers to develop their own self-learning environment. It is the individual or the company they work for, which tends to identify the need for learning, development, progression, or training in order to comply with the law. It is the law which is driving learning, rather than any pressure from consumers or the insurance industry responding to past tragedies at events.

Because individuals or the company they work for is driving the impetus for education and professionalism in the industry, there is an autonomous and fragmented approach to self-learning. This has impacts for elements such as setting industry KPIs and raising the need for the production and reproduction of knowledge in the industry without an international professional association leading the way for organisers to align their behaviours and practices across the marketplace for international events.

3   Tourism and Events

This chapter is concerned with how event organisers in the developed and developing constructs approach the tourist sector of international events. 'Event tourists' incorporates international clients who book venues abroad, and foreign consumers who travel to attend an event. The theme of tourism and events, therefore, is a core element in the context of gaps within the global marketplace for international events.

The organisers who were interviewed each find themselves in direct engagement with event tourists by organising events which attract international consumers (Ferdinand and Kitchin, 2017) whether clients or attendees. This is because they are organisers at landmark capital city venues which of course attract international events.

It is recognised by Getz and Page (2016) that tourism is closely related with events because of the importance of planned events as 'products' or 'attractions' and that events are consumed in the leisure time of tourists. Ferdinand and Kitchin (2017) agree by describing the international approach to events management as sharing a focus on other international activities such as tourism and are concerned with differences between cultures and countries. They argue that an international approach to events management requires engagement with international activities, cultural differences, and global issues. So, tourism in events is not only about clients and consumers, but it can also be about organisers travelling in their job role:

> *I moved to New Zealand to work on the America's Cup, then I came back to work on the Olympics in Athens. Then I worked for Ellen MacArthur around China. Then I moved out of sport because I got stuck in a rut. Then, I went back for the 2012 Olympics.*
>
> (UK9 – organiser at an institution)

This excerpt demonstrates the existence of a dual approach to event tourism where this organiser performs the role of a tourist searching for new experiences, for something different (O'Donnell, 1997), while undertaking their job organising events at those host locations and experiencing the needs and

DOI: 10.4324/9781032684161-3

expectations of leisure tourists, and what attracts tourists to unique events (Gelder and Robinson, 2009). The internationalisation of events allows this to happen and is indeed one of the motivators for people choosing to enter the events industry, as has long been with the closely related industries of hotel, tourism, and aviation offering travel opportunities, international experience, and worldwide networking.

Event organisers, like most people, would claim they know what it is like to be a tourist, because of their experiences of travel in their leisure time. But the above organiser is highlighting the potential to place oneself in the position of gaining experience as a 'working tourist' and gain experience of organising events for tourists, while being a tourist. Many young people achieve similar by taking a working 'gap year' and can bring their international experiences into their early career. This type of tourism is helpful in narrowing gaps between developed events industries and those which are emerging, because travelling organisers will export and import knowledge, experience, and practices.

However, outside expertise is observed to be routinely rejected in the emerging events industry in Poland. This fact draws a distinction between leisure tourists who are visiting only for pleasure and are welcomed purely for their spend, and professional tourists who bring with them their own ideals drawn from experience and expertise, which is not welcome in a socio-professional environment that is already practising its own way of doing things, rightly or wrongly. The rejection of expertise from abroad due to sociocultural differences is a factor that will impede the advancement of a developing events industry and slow the opportunity for narrowing gaps in the marketplace for international events.

3.1 Event Tourism

For economic benefits and job creation, governments have turned to tourism as a growth industry, with events seen as image-makers and creating positive profile of a destination. It is recognised that event-goers will spend on travel, accommodation, and other goods and services in the host city (Bowdin *et al.*, 2011). Destinations have realised this impact from events and are developing and implementing strategies to increase visitation (Bowdin *et al.*, 2011). Locations are spearheading event tourism as a social phenomenon by developing events and the infrastructure for events to attract consumers who would be motivated to travel to an event.

According to the World Travel and Tourism Council, tourism is the largest industry in the world and is growing at an extremely rapid rate (O'Donnell, 1997). So the value of tourism and its growth is not disputed. Event tourism can be considered a hybrid genre of events designed to attract consumers to travel to an event in 'the development and marketing of events for tourism and economic development purposes' (Getz, 2008, p. 406, cited in Ferdinand and Kitchin, 2017, p. 318).

Hays, Page and Buhalis (2013) identify social media as the tool to reach a global audience with limited resources in a rapidly changing culture because it reaches people at a scale and speed larger and more quickly than previous communication channels. To contextualise the growth in usage of social media, Qualman (2009, cited in Hays, Page and Buhalis, 2013, p. 230) reports that if Facebook were a country it would be the fourth most populous in the world.

In recent years there has been growth in places to where consumers travel to experience certain features, characteristics, or a perceived attraction of some sort (Buhalis, 2000). In Poland, cultural consumption is shifting from high culture (museums, art galleries, religious sites, historical landmarks) to popular culture (festivals, sport events, performing arts attractions) (Marciszewska, no date, cited in Smith and Robinson, 2009). It is therefore important for event organisers in city centre and capital city locations to reach and penetrate the foreign event tourism market by developing integrated services to meet the needs of tourists and remain competitive (McCabe, Sharples and Foster, 2012):

American clients demand a quick response – four hours is a typical turnaround for a proposal. I use Microsoft Teams and a show-round app to [convey] *visuals. I visit countries to see their culture and build relationships with long-term planners.*

(UK7 – landmark conference centre)

This organiser employs technology to facilitate the process and enact speed to foreign clients who *demand a quick response*. Driven by their astuteness of proactive commercial acumen, such as using *a show-round app to* [convey] *visuals*, it is reasonable to accept that this organiser is driven by broader social, cultural, and economic changes in the event tourism sector.

3.2 The Trend of Tourism

Industrialisation and increasing commodification in the tourism sector brought about the desire for escape and distraction, explain Wearing, Stevenson and Young (2010). They correlate industrialisation with the administration of time, because more precise clocks drove the measurement of time through timesheets, overtime, timekeeping, and time-off. This led to leisure being taken in commodified time or 'sacred' time (Wearing, Stevenson and Young, 2010, p. 37), because leisure time is a space for gaining social capital which is sacramental to human existence (Kielbasiewicz-Drozdowska, 2005).

Poland is not immune to changes in the leisure tourism sector and the opportunities this offers. According to Kielbasiewicz-Drozdowska (2005), social changes in Poland have been strongly connected with political and economic transformations and with changing social awareness. She recognises that profound changes in the workplace have led to re-evaluation of the way in which leisure time and work time are understood. This cements the global impact which international travel has on all locations. But event organisers in emergent Poland are not yet capitalising on capturing the event tourism market and can learn from the proactive, immersive, and technological approaches adopted by organisers in the developed UK industry.

Buhalis (2000) recognises destination marketing is increasingly competitive worldwide and requires taking advantage of new technologies to enable destinations to increase their visibility. Further reinforcement of this point is provided by Salameh (2019), who writes that the FIFA World Cup and the Summer Olympics has been the single most important way to reposition the international profile for Brazil, and Giulianotti *et al.* (2015, p. 99) refer to the London 2012 Olympics as the opportunity for 'festival capitalism'. Gillespie and Hennessey (2016) concur that competitiveness moves across national borders at an alarmingly fast rate.

The fast approach reflects the recognition by Hospers (2004) that European regions are increasingly competing to attract consumers through soft factors such as a location's image, referred to as 'place marketing'. Warnaby and Medway (2013, p. 345) describe this as 'the marketing of towns and cities . . . to gain advantage for their particular place in an increasingly competitive spatial environment'. They further categorise place marketing as creating offers commodified and marketed to target consumers through the attraction of tourism and a sense of place-attachment.

3.3 Penetrating Foreign Markets

UK organisers are bidding for events from large foreign clientele to landmark venues in London's intensely competitive marketplace with its range of large conferencing venues making it a bidding market at the stage where the competitive contract to run an event is won or lost (Berridge, 2010, cited in Berridge, 2010). Quinn (2013, p. 8) describes this as a 'very risky activity'. Conference News (2019) confirms that London continues to be a hotspot for North American meeting and events planners with bookings in 2018 up 20 percent on the previous year. Large international clients will place their enquiry to several London-centric venues and the winner could be the venue which responds quickest. To meet this demand, the following organiser not only waits for inbound enquiries but makes business trips to their client locations in the United States:

> *I attend the PCMA* which is an event for event professionals. There are 4,000 delegates for educational, networking, sales and sharing best practices.*

(UK7 – landmark conference centre)

---

* The PCMA (Professional Convention Management Association) (2020) describes itself as a professional resource for leaders in the meetings, conventions, events and trade show industries with over 6,000 members to facilitate connections.

Attending stateside conventions for networking in the home market of target clientele is a proactive approach whereby the organiser is travelling abroad to secure foreign business and counteracting the very risky (Quinn, 2013) bidding process. This demonstrates a plurality of approaches to destination marketing: the location contriving an image to attract visitors from foreign markets, and a marketeer reaching into foreign markets to draw visitors to the destination.

Proactive engagement activities with foreign markets such as place marketing (Warnaby and Medway, 2013) and place branding (Kavaratzis, 2004) contradict observations of Poland organisers in the emerging industry. Whereas UK organisers invest time and financial resources to travel abroad to client destinations to secure international business, there is a gap because Poland organisers demonstrate a reactive or non-interventionist approach:

They [the Polish Convention Bureau] *are sending us some requests for venue leasing, and somehow they are promoting us as a place amongst foreign customers. I would say they do marketing, mainly marketing. We have got some associations – MICE associations mainly hotels and that kind of venues we're working with them – somehow they promote us amongst their international customers, somehow.*

(PL2 – landmark venue)

In this excerpt, the interviewed organiser wavers and uses phrases such as *somehow* and *how to say this diplomatically*, which indicates they do not fully understand how the Polish Convention Bureau operates and how they should be engaging with them to promote their prominent landmark venue to foreign clients. This organiser instead waits for foreign business to arrive, should it happen, without being actively engaged in pursuing or securing foreign business for the venue. This contrasts with organisers in London where the competition is fierce, and they are driven to actively penetrate foreign markets. This factual is not to identify complacency but to highlight this gap in the socio-professional behaviours between the developed and developing industries.

For sure, Poland organisers do possess the will to drive their events business into the global market for international events:

Huh . . . I would say that I would wish myself to be able to put more attention to promote the building and to promote the support internationally. I would say that it is underestimated – our uniqueness of the building. But I am also aware that there are so many challenges organising an event here. So, I would say that firstly we should fix those technical and regular issues and make things more commercially competitive to be able to promote the building internationally and to have more international customers. I would say that we should put more attention to that. Yeah. And I would say this should be our priority as event managers. Not only do our daily job to focus on technical issues and signing contracts, but to do more PR and marketing internationally.

(PL2 – landmark venue)

This is encouraging because although it does show the lack of know-how – which is symptomatic of the limited community of practice in the shape of an emerging events industry – know-how can be gained from international networking and knowledge transfer and addressing the gaps which exist between the developed industry and the emerging industry.

However, the above organiser is at a state-owned property and feels unsupported in their ability to reach international markets. It is a situation which resonates with the finding by Wong, Mistilis and Dwyer (2011, cited in McCabe, Sharples and Foster, 2012) that barriers to collaboration include a lack of coordination among government departments. This is evident with this organiser who is responsible for the day-to-day running of a state-owned venue, such as dealing with *technical issues and signing contracts* but recognises there is no strategic forethought to win business from foreign markets. The Chartered Institute of Marketing defines this as 'the management process responsible for identifying, anticipating and satisfying customer requirements profitably' (*Chartered Institute of Marketing | CIM, 2021*).

This organiser is in a situation identified by Welch, Benito and Petersen (2018) that international managers rarely face foreign market entry decisions because of the daily concerns of their job such as growing the business, reorganising, and developing existing operations. That is the limitation described by the above organiser: *I would wish myself to be able to put more attention to promote the building and to promote the support internationally.* It is a restriction for them in reaching out to foreign markets because they are kept busy with the operational demands of running a landmark venue. It is not a situation relevant only to Poland because this UK organiser exhibits the same focus on daily operations:

My main role is to facilitate our organisers' needs to get as close to what people want. We'll always try and help . . . to achieve all of this. That's our main function, really. We take good money off clients, so we have to give them good customer service. I get very annoyed when we don't deliver good customer service. I think keeping continuity and a sustainable business arm running – we try and run an efficient and lean team and try to be a bit more streamlined, I suppose, but very much keeping an eye on efficiency levels making sure we are doing all we can. We've spent a lot of time and money on training staff for the purposes of serving the business as opposed to personal development. It's always been very much based on what our output is.

(UK10 – landmark exhibition centre)

The focus of this organiser is firmly on business objectives and the sustainability of the business rather than on consumer objectives and meeting expectations of consumers. Saying *we've spent a lot of time and money on training staff for the purposes of serving the business as opposed to personal development* shows their focus is not on developing people but enabling them to do better purely for the needs of the business. This is a business-centric approach at the expense of people development or even recognition of consumer needs and expectations. Conversely, the following organiser does focus on consumers, but this is still for the purpose to serve the business's objectives:

People would like more value for money. I feel that the spend per head has increased but they're expecting a lot more with that. I feel that they are using the [venue name] *for special occasions. Predominantly, a lot of guests here it's potentially their first time here at* [venue name] *and it's normally a special occasion – it might be their favourite act that they've waited their whole life to see. We've noticed that it's not just becoming one event per year, a lot of people do return to see different events. I think it varies between people who come to* [venue name] *to see a lot of acts and some that still come when* [artist name] *decides to do a show here and they've waited their whole life to see her and now they're coming to see her for the first time. But I think everyone has that idea that they spend more now.*

(UK17 – landmark arena venue)

These excerpts reveal that it is not the type of objectives an organiser focuses on, which is going to impact their reach to event tourists (whether they are foreign clients or foreign consumers) but whether they are focused on operational level or strategic objectives. Both the above organisers focus on different operational factors, and the former organiser is stuck on them, and each is not going to attract event tourism this way.

There is a difference between the UK and Poland in that UK organisers have the freedom to refocus their objectives, but Poland organisers do not. Yet it is the Poland organisers who want to refocus their objectives from daily operational issues to strategic marketing to reach foreign markets.

There exists an assumption that organisers in the developed industry would be better equipped to reach foreign markets because theirs is a multi-layered collaborative structure of organisers, marketing managers, and public relations people, whereas in the emerging industry, organisers are likely to do everything themselves (which they cannot do) because it is a flatter, less structured industry. However, this problem exists in both territories. Thus, unique landmark venues in any capital city should be positioned for bidding in the global marketplace for international events to attract international clients and foreign consumers to that destination.

Although Buhalis (2000) views destinations as amalgams of tourism products offering an integrated experience to consumers, the following organiser reveals how landmark venues are not attempting to win individual tourists, but those clients who bring the tourists:

> *At least 85% of the time business comes from event management agencies or venue finding businesses. Most of the big companies are mandated to use event management agencies or venue finding businesses. With that in mind, you need to ensure that you don't just have the relationship with the end client, you must have the relationship with those third parties as well.*

(UK13 – sales account director)

Event agencies and venue finders are less abundant than individual tourists, so there is a narrowing to the shape of the structure which means the competition to win that business gets fiercer. Wong, Mistilis and Dwyer (2011) note the conflict between cooperation and competition, but there is a lack of competitiveness in the Warsaw marketplace because it is an emerging location with less demand than London. This is working against venues in Warsaw making efforts to win foreign business and promote themselves and the destination to the global market for international events – because they do not have to.

Venues in locations with intense competition are driving organisers to reach foreign markets for business and promote the destination. Gillespie and Hennessey (2016) identify developed economies like the UK account for a disproportionately large share of world gross national product. As a result, UK organisers need to attract many companies into their market environment:

> *The credit crunch hit in the backend of 2008, and this is where the type of venue is very important.* [Venue name] *had an incredible level of reoffer business. What that means is that a function will happen on the same day every single year, notwithstanding – insurance awards, banking awards, actuary awards will happen on the same day every single year. What people did with the credit crunch is they said, 'we're not going ahead with our function, but we will retain the date and we will pay you not to sell that date and we will recoup it the next year'. So, we went into negotiation with several clients and were charging somewhere around 60%*

cancellation fees and that represented the profit we would have made on their function.

(UK15 – landmark event hotel)

3.3.1 Response to Trends

The above excerpt shows how events can adapt to reap a competitive advantage. Not many industries have this ability to be flexible so rapidly. Hotel bedrooms, for example, cannot respond easily and quickly to trends or financial climates. This is important because events generate income particularly in times of uncertainty when competition for business intensifies – such as happened with the critical impacts of the COVID-19 pandemic:

The budgets had decreased pre-COVID which has actually changed quite dramatically post-COVID. Whereas now I think there are far less events, but the budgets are far higher, I think prior to COVID it was just more and more and more events and the budgets were smaller and smaller and smaller. In the UK you get a tax break on your Christmas party per person and we were seeing a massive shift in companies splitting that budget down the middle and using half of it for a summer party and half of it for a Christmas party which was only really, I would say, since about 2015 when we started to get better summers but prior to then when we didn't have any nice weather there wasn't any point. I would say the early 2000s budgets were far higher for events because nobody was really checking. And then 2010 onwards there was a bit more of a check on those budgets but there was [sic] more events and then, yeah, we've kind of come out of COVID in a slightly different direction again. And whenever something new comes out – a new style of event – like when molecular food came out and like you know fancy fine dining that was a different burst and like I say about summer parties that was a different sort of style of thing now obviously we're seeing hybrid and virtual events sort of taking off in a way that I would never have imagined.

(UK14 – portfolio of royal museum venues)

This organiser affirms that events are adaptable to sudden changes by saying *it has gone up and down in different ways with different themes.* It is this rapid adaptability which organisers need to exploit with their event offering and it seems organisers in the developed structure are aware of this and do adjust. This is an essential trait for the sustainability of an events business in a culture of an unstable financial climate, rapid social and technological changes, and in times of uncertainty.

Organisers in the shaping structure are slower to identify trends and less likely to adjust their event offering. This is because they are still in the learning phase and will be focused on other areas of learning, and they are working in a less joined-up structure to learn trends. There is a gap between UK and Poland organisers in recognising that events have the specific characteristics to be able to rapidly adapt to change and win business from competitors in the marketplace. This puts the Poland events industry at a disadvantage to capitalise on growth from the wider global marketplace:

> [Events] *are hugely important. Especially for here at* [venue name] *our fine dining restaurant is quite expensive to run, so without bringing in the money that events brings in, it would run but it wouldn't have as strong a staff and team because, you know, whatever we bring in through the events comes through* [venue name] *as well so it's really important for bringing that huge amounts of money for the business.*
>
> (UK16 – multi-purpose venues)

This organiser in the developed industry shows their understanding that events are the leader in their business, and the revenue contribution from events allows other areas of the business to keep going – areas which may be more costly to operate or achieve less profit in proportion to costs. It is important for venues to win events business from competitors to support other areas such as maintenance and upkeep, marketing, penetration into foreign markets, and investment in technology. This is especially important for a diverse business but reinforces the significance of events for any business:

> [Events] *sits very high up. If events are busy it will have a dramatic impact on the accommodation level of occupancy. If events are dead, the occupancy of the hotel will stem anywhere between 30% and 50%. In September, October, November and the first half of December we* [events] *were chipping that up to around 80% to 100% so we have a dramatic impact on the level of business the hotel takes.*
>
> (UK15 – landmark event hotel)

This organiser relies on events business to raise and sustain other areas of operation – in this case, the accommodation occupancy levels. One might have believed that a landmark hotel in central London would be full of guests and its primary source of income would be from bedrooms. But in fact the accommodation side of this business is sustained by events – the occupancy levels are wholly influenced by events business in this hotel.

UK organisers recognise the importance of events in a competitive marketplace. Ohmae (1985, cited in Gillespie and Hennessey, 2016) highlights the importance of developing a competitive position in major developed markets – in particular to compete in the United States, Europe, and Japan, where 'real global competitors are advised to have strong positions in all three areas' (Gillespie and Hennessey, 2016, p. 249). But Poland organisers are not attuned to competing in the global marketplace of international events or spotting global market trends.

The growth effect of an industry is to first evolve the domestic market before reaching abroad to overseas markets. In an evolving market, organisers are occupied with learning the craft of managing events, let alone reaching for foreign clients and attendees – or knowing how to do so.

Another difference observed in the developed industry is that UK organisers are not only active in reaching foreign markets, but they are also attracting event tourism business in more than one way. Organisers do not necessarily need to visit foreign markets to attract clients and attendees but can participate with events taking place abroad:

> *We did* Global Rhythms *working with the Trinidad and Tobago Commission – they very much wanted to profile their location and their country for tourism. It was working with the tourist board bringing a little bit of Trinidad and Tobago to England and we managed to bring across a number of high-profile artists but the challenges around that was the expectation once they're here. Dealing with an international artist who may have a rider* who is an A-list celebrity in their own right in their own country, bring them over here in terms of having a rider which is suitable to all was quite a challenge as to what their expectations were and how their fees are and when and how those fees should be paid. A lot of international artists work on the proviso that they get paid when they're here rather than having a pre-payment. That was an attendance of close-on 8,000 to an event like that. We also had the Commissioner from Trinidad and Tobago coming to the event, and high-profile UK celebrities.*
>
> (UK5 – organiser of local authority events)

What this organiser describes is different from the usual promotion activities of a destination putting on events to attract foreign consumers to attend at the host destination which serves to promote and enhance that location and enjoy additional benefits such as tourist spend. Here, this organiser demonstrates how a destination can export its location through an event taking place at a foreign location. In this case, instead of Trinidad and Tobago attempting to host events that would attract tourists to travel to attend on Trinidad and Tobago, they brought *a little bit of Trinidad and Tobago to England* to showcase their location. Urry (2002a) identifies this as object travel, which is the object of attention that draws the tourist – in this case it is whatever Trinidad and Tobago are promoting such as culture; lifestyle; or simply sand, sun, and sea (Urry, 2002b, p. 51).

It raises the question whether an organiser in an emerging market will recognise that they could promote their location and their venue by exporting

---

* A 'rider' lists the contractual requirements of an artist or a performer and can be technical requirements for their performance, or hospitality requirements for their comfort.

it to an event taking place in a foreign location. Such an event might be a tourism exhibition or an international venue convention such as International Confex at the ExCeL, London (*International Confex – Where the events industry meets*, 2021).

However, Welch, Benito and Petersen (2018) caution that reaching foreign markets is not the end of it and there is life after entry. Their view is that initial entry to a foreign market is a basis for building business and will need to evolve with changes of operation. This type of engaging with an industry abroad is an opportunity to win business, promote the venue, facilitate knowledge exchange, and understand cultural differences and the feeling of industry inclusion.

The developed stage of the UK events industry means that there are more opportunities for organisers to reach foreign markets because international associations with affiliated groups in the UK (Bowdin *et al.*, 2011) provide a route of entry and exposure:

> *I went to the AIPC* [International Association of Convention Centres] *out to Brussels for a week and we discussed things such as international delegates coming to our venue and international clients coming to our venue. Being part of international associations, we have exposure to meetings, conferences and seminars. We talk a lot about working with international clients and how we as UK venues set best practice and how they work across Europe or the Middle East or North America.*
>
> (UK6 – landmark convention centre)

Clients, event agents, and venue finders abroad are the core target market for this organiser because of the significant size and central London location of their venue. Here, the organiser demonstrates their activities in foreign markets and international collaborations with the objective to draw foreign business to their venue and, in this case, develop an understanding of the expectations of *international clients coming to our venue*. There is a clear distinction being identified here between the shape of the socio-professional UK habitus and that of emergent Poland: the difference of active penetration into foreign markets versus the non-interventionist.

Placing oneself into an unfamiliar field (or 'habitus') takes time and adjustment to fulfil what Bourdieu (no date, cited in Grenfell, 2012) identifies as the need to match the habitus with a social context so as to feel comfortable and not avoid or resist participation in that social context. It is about acceptance because an outsider will not sit comfortably in the habitus until they accept, and be accepted, as existing in that social environment.

Bourdieu describes fitting-in with the habitus as 'practical mastery' or 'feel for the game' (Grenfell, 2012, p. 57) (but might otherwise be described as 'engaging with industry networks') to understand what is reasonable and unreasonable, likely and unlikely, and natural ways of taking actions as a mediated form of arbitrary social structure.

3.3.2 Internationalisation

Event tourism is not limited to clients and attendees. It extends to organisers themselves who can be attracted to foreign destinations to place their event:

> *Some institutions in Poland are absolutely on the international level and they are highly professional people. But on the whole if you visit a smaller event somewhere they do adhere to all the regulations set up by the EU. I think joining the EU was a very, very good thing because you have certain regulations which absolutely have to be seen to. But sometimes I do come across things which for us would be unthinkable; poorly managed; poorly organised. But some are absolutely on the international level, I would say.*
>
> (PL4 – organiser of charity events)

This organiser places their events in Vienna and is discussing standards in Poland in the context of comparing the two locations. By doing so, the organiser demonstrates the acquisition of knowledge from their wider experience of international standards and practices. But the point made about EU regulations is pertinent: *I think joining the EU was a very, very good thing because you have certain regulations which absolutely have to be seen to.* An emerging event industry will evolve and take shape in its own manner according to its cultural and societal environment. But when Poland joined the European bloc in 2004, it adopted EU-wide legislations and regulations which continued the drive for Poland to align with the standards and structures of other European countries (Domański, 2003). Poland's accession to Europe pivoted the Poland events industry to fall in line with international standards and expectations in catching up to the advanced economies and competing successfully in the global market (Bogdan *et al.*, 2015).

Observations concur that domestic events in Poland can be substandard because of the lack of local professional expertise in the management of events. Emerging locations win 'mega-events', which bring with them infrastructure, guidance, and outside expertise. Thereafter, that location has 'emerged' into the space for international events, albeit without local expertise after the mega-event has ended.

But the above organiser is identifying how European regulations are improving standards of events in Poland. Indeed, Gillespie and Hennessey (2016) note that there are moves to internationalise laws, with the Hague Conference on International Private Law considering a global treaty for enforcing legal judgements. This would be a significant move in the direction of standardisation and would help destinations to become associated with desirable qualities perceived by target audiences (Kavaratzis, 2004).

It is an argument for homogenisation, which would narrow some gaps that exist from location to location. It would no longer be the haphazard shaping of an event industry according to its own socio-professional environment but a standardised homogenous international event industry led by those countries which are already shaped and developed into a socio-professional structure with all it entails. A move such as this would recognise the theory advanced by Mary Douglas (Anfara and Mertz, 2011) in identifying four prototypes of social environments which influence the entire cultural environment. These are individualist (individuals not constrained by rules or traditions), bureaucratic (hierarchical with low levels of autonomy), corporate (individual behaviour is exercised in the name of the group), and collectivist (the perpetuation of group goals and survival is highly valued).

Bogdan *et al.* (2015) appraise that the initiative which the Polish people have shown indicates that they aspire to become competitive on a global scale. The above organiser is evidence of this appraisal because they are in the process of comparing domestic standards in Poland with standards encountered in Vienna. This organiser is gaining learning of international standards which makes them a better organiser. If this were upscaled and became embedded in the socio-professional behaviour of organisers across the events industry in Poland – if other organisers in Poland achieved learning through international experience – Poland would narrow their gaps in the global market for international events.

The following organiser reinforces this potential by their comparisons of different locations drawn from experience in the Middle East and Malaysia:

> *Certainly when I was overseas you wouldn't really bring up deposits and payments and schedules and that sort of stuff if you were dealing with Middle Eastern clients – it was considered to be uncouth. In Malaysia I remember taking an enquiry for a French event management company and they expected us to be quite European in our attitude and I had to tell them upfront that things didn't happen perhaps the way it would happen if you were in Western Europe, out in Malaysia. If you've got a Western European or American agent wanting to bring a piece of business, they want reassurance that things are going to happen to time and to budget and all those normal things you'd expect. But you also have to share the realities of the part of the world you are in.*
>
> (UK3 – landmark sports venue)

It can be seen from both these organisers how international experience is a valuable contribution to the profession in both the developed and developing event industries. The latter organiser understands that it is cultural differences which shape the way things are done and not only about knowing how to organise events. In fact, they are suggesting that knowledge in organising events is secondary to getting the cultural customs correct by saying, *you also have to share the realities of the part of the world you are in*. This understanding was learnt from their interactions in foreign fields which otherwise might not have been understood.

Gaining such understandings fits with the four characteristics of profession identified by Ritzer and Walczak (1986) as theoretical knowledge, self-regulated training and practice, authority over clients, and orientation to community rather than self-identify. Theoretical knowledge can be obtained through regular interaction with peers, and authority over clients provides organisers with knowledge their clients lack, which allows them to provide direction and advice when responding to clients' demands. This is made clear by Berners (2019, p. 67), who states that 'above anything else a client is looking for guidance and support – otherwise they would be doing the job themselves'.

Clients rely on guidance from specialists, and the specialists rely on guidance from other specialists – otherwise the system fails (Berners, 2019). The basis of being an event organiser is to advise others – a concept referred to as the 'real' event manager (Berners, 2017, pp. 19–21). The 'real event manager' concept is compatible with Foucault's power-knowledge theory (Gutting, 2005, p. 96), investigating the traits of seriousness and credibility: 'who was authorised to speak seriously, and what questions and procedures were relevant to assess the credibility of those statements that were taken seriously'.

So, organisers in the UK events industry benefit from their penetration into foreign markets which is supported at local and national government levels because of the significant contribution events provide to the economy. Events in Poland are not yet recognised with that same level of seriousness and Poland organisers are not supported in their reach and penetration of the foreign event tourism market. By not doing so, organisers in the emerging industry are not gaining the knowledge transfer brought about with international engagement and relations, they are not spotting and meeting global trends and demands in events, and they are not being driven to operate at the level of standards recognised in the global market for international events.

3.4 The Experience Economy

This book is concerned with the shape of two events industries within the global market for international events, so it is important to address the experience economy because events are live experiences designed to attract consumers and clients in the global market of event tourism.

The experience economy is about making money out of experiences (Boswijk, Thijssen and Peelen, 2007). Pine and Gilmore (2011) provide the example of a coffee bean to explain how the cost of a service can be inflated by the experience it delivers to the consumer. The cost of producing coffee from a coffee bean is far less than the purchase price of a coffee in a coffee shop. A consumer can choose to further inflate the purchase price for their coffee if they decide to purchase it in, say, Harrods or The Ritz. They will select to pay a higher inflated price just for the experience of consuming a coffee in that environment, which is how the experience economy works. Gerritsen and Van Olderen (2014) consider events to be typical of the experience economy because they meet the demand for unique, personalised, and memorable experiences and provide social cohesion in communities. Consumers will pay a premium to attend an event for the experience it will give them which they will not get anywhere else or if they do not pay to attend that event. Thus, creative events are typical of the experience economy:

> *For me it's always about people leaving in a happy fashion, you know, they've achieved what they needed to do. You see people leaving – you only need to hang around in an exit area looking kind-of official and people are very quick to tell you how they feel and think. I've worked with other stakeholders and done lots of concerts and music events with 20,000 people piling out all at once from* [name of venue] *and you get the Tube station manager saying, 'thanks very much it went really well' and police officers saying, 'thanks very much it all went well'. And job-well-done really – that's what we're there for; we're there to make our partners' lives easier so they're doing their jobs safely and efficiently and very much with a customer focus, visitor focus in mind.*
>
> (UK10 – landmark exhibition centre)

This organiser describes the experience economy: creating an experience for the consumer who has decided to participate by paying to attend. For this organiser, their job is to provide the experience their consumers have paid for and expect to receive – this is how the organiser measures the *job-well-done*. This organiser is passionate about their job of delivering an experience to the consumer. By doing so, the consumer will receive a transcendent experience that would otherwise not be achievable (Rojek, 2013).

It is therefore important for event organisers to understand the motivations of consumers which, for events, is socialisation, a fun atmosphere which offers ample opportunity to socialise, and to have new experiences (Gelder and Robinson, 2009).

If the potential consumer decides not to participate, they risk missing the opportunity of receiving the experience because the event will be gone. This is what makes events a perishable service (Bowdin *et al.*, 2011) because when the date of the event passes, it cannot rehappen or be resold. Organisers in the developed UK events industry recognise they exist within the culture of an experience economy:

> *There is far greater emphasis on an experience, which works very, very well where I currently work but less well for people who don't have an*

experience to offer. So, I think clients want to have something that they can take a picture of and something that they can remember. So, for example the museums, I can offer things like a telescope viewing or a planetarium show, or a gallery tour, or all of these other sort of added value things which a hotel can't because it's a hotel. So, you know, anything that you can do to make it seem a little bit 'sparklier' a lot of clients are asking for sort of what else can we do to enhance people, potentially because there's more events for people to go to, they want people to come to theirs.

(UK14 – portfolio of royal museum venues)

This excerpt highlights the experience economy in the developed events industry has already elevated to an ever-upward requirement for experiences. Driven by demand for more exciting events and creativity, experiential events have shaped the events industry in the UK, because at one time creating an experience used to be an additive to an event, but now experiences *are* the event.

The above organiser recognises the demand for event experiences and understands the competitive advantage of their venues meeting that market. In fact, this organiser mentions hotels in almost a derisory fashion: *[A] hotel can't because it's a hotel.*

3.4.1 Hotels' Loss of Events Business

Creativity for an event can be a challenge if it is in a hotel ballroom, which is typically 'designed to be generic' and 'do not allow for the creativity of events' (Berners, 2019, pp. 11–12). There are exceptions where a hotel event space is architecturally impressive, or where the hotel ballroom is exceptionally large such as the Grosvenor House Hotel on London's Park Lane which can seat 2,000 guests for a sit-down banquet, or when the event requires overnight accommodation for guests, in which case it must be a hotel.

Hotels in the developed events industry have lost significant share of events business as the industry expanded to other venues that became competitors in the events marketplace and are more exciting, unusual, and different – and better aligned to fit the experience economy. As the demand for events grows in a location, it demands ever more creativity, excitement, and experiences. This causes events to 'leak' from happening only in hotels to events being staged in all manner of venues, from art galleries and museums to disused warehouses. This has a cause-and-effect impact because when events move from hotels, they are more complex and require specialist expertise, knowledge, and resources. In turn, organisers need professionalisation, skills, and experience. And, it can be charged for, so it means higher event budgets and larger profits.

Observing the emerging industry shows that events are more inclined to take place in hotels, underpinned by Buczak *et al.* (2020) reporting just

9 percent of events in Poland are hosted in congress centres and only 16 percent in event venues, compared to 65 percent hosted in hotels. Events in hotels are less risky because they have resources *in situ*, including knowledge and skills in hotel management as a related discipline. Still, the events industry in Poland is not as confined to hotels to the extent the UK events industry was when it was evolving. At one time in the UK, events were 'hidden' in hotels and visible mainly to hotel workers. This restrictive factor influenced the shape of the UK events industry as it is today because current senior-level organisers crossed into events from their roles in hotels, such as event sales or conference and banqueting. Nowadays, organisers do not need to work in hotels to work in events, which has helped events management to identify as a stand-alone discipline. It has also driven the rise in formal education in events management because it is no longer seen as an offshoot of the hotel discipline and managing a hotel department.

The reason the emerging industry is not as confined to hotels is due to the cross-cultural and global context of events which were less so at the time when the UK industry was evolving. Event clients and consumers are more widely travelled and have wider experiences of events at home and abroad – especially as event tourists. The events industry in Poland, therefore, has leap-frogged the years of events being confined to hotels.

There are two further factors why hotels have lost events business. The first is the nondescript design of function suites designed to appeal to any types of events from formal conferences to creative parties. In many cases, such function suites in hotels have low ceilings because of cost restraints during construction and architects being ignorant of the needs of events. Events require high ceilings for the creative elements (Berners, 2019) and it is difficult for an organiser to insert creative elements without generous ceiling height. Organisers and event suppliers (and clients, actually) are judged on the creative experience of their events – including feedback from attendees – so an event must be creative. Plus, organisers and suppliers make money from providing the creative elements to enhance the event experience. So, if there is no height in the venue for creative enhancements, organisers and suppliers will place their clients elsewhere into venues with higher ceilings.

Second, there is a perception in the UK events industry that hotels are for low-budget events. This has come about because hotels provide most resources *in situ*, including catering, staff, and notably technical apparatus which is less complex and less costly. So, if an event budget is not generous, a hotel would be the best place for that event. This has resulted in negative connotations for clients in the developed industry who have reached the understanding that if their event takes place in a hotel, it might be seen as a low-budget event.

This shift is still happening and relevant because, as the above organiser states, their venues are unusual, interesting, and creative, which allows for events held there to be individual and memorable: *There is far greater emphasis on an experience, which works very, very well where I currently work.* This is an important competitive advantage to win business from hotels in an experience economy because the events offering is unique and can be tailored to clients looking for evermore creative experiences and *anything that you can do to make it seem a little bit sparklier.*

The upward demand for creativity and uniqueness is driving UK venues to adapt their events offering to capture clients in the experience economy marketplace. In fact, the following organiser speaks of the need for their hotel to pivot from corporate events and move into the creative events space purely to align with the experience economy:

> *It involves change and the growth of our business. We, as an industry have to adapt. We, as a business – one of the plans for our 10-year anniversary is to look at expanding the function space and changing that. We have to develop it as an event venue as opposed to a meeting venue – it's very corporate at the moment. It's very functional – it's very adaptable, actually – but we have to lean more towards events than meetings when we develop that area. We need to expand the space because we need bigger numbers . . . and in terms of the interior design of it needs to be more in keeping with events, including weddings, than a board meeting or a conference.*
>
> (UK11 – country house hotel)

This organiser underpins the problem with generic hotel function suites by saying, *It's very functional – it's very adaptable, actually.* It sounds a strength, but it is no longer enough for an event space to be functional and adaptable when the demand is for unique creative experiences. This is a significant shift of this hotel's business mix from the existing corporate events offering to that of creative events. This reaffirms the shift of the UK events industry and confirms that venues are realigning in response to the experience economy.

Poland organisers also appear to be recognising the experience economy:

> *Most of our customers, they come back. Most of them. But, I'm aware that not because of the quality of our job but because of the building – it's unique. That's the reason. It's a really good spot for B2C* [Business-to-Consumer] *events, for fairs and exhibitions to the B2C market. The best one, I would say here in Warsaw. That's why customers come back to us – not because of the quality of our service, yeah our uniqueness makes the job.*
>
> (PL2 – landmark venue)

This statement is clear indication of awareness in the emerging construct for the need to provide an event experience. The above organiser demonstrates that event consumers in Poland are looking for uniqueness, just as in the UK.

However, unlike the UK, organisers in Poland have not yet begun strategically aligning to the experience economy.

Poland, therefore, again demonstrates a non-interventionalist standpoint, as with their penetration into foreign event tourist markets. Moreover, the above organiser is candid about consumers not choosing this venue because of the quality of service they will receive but because it is a venue which offers a unique event experience. This organiser, then, is banking on winning event business for the uniqueness of the venue, rather than having to provide a quality service to win that market sector.

This exposes a gap between the UK and Poland in meeting the experience economy because while both locations recognise the value of providing an experience for event consumers, organisers in the UK are actively aligning to that market sector. Poland organisers, however, are content to let that market arrive to them without making efforts to win that sector through market reach and penetration, or even to meet the service needs and expectations of that market once it has landed.

3.5 Gap in Service Quality

When the above organiser states, *I'm aware that not because of the quality of our job but because of the building – it's unique,* they are accepting their shortfall in providing quality service. This is another indication of an experience economy in Poland because it reveals clients are placing quality of experience above the need for quality of service. The problem is that it exposes a gap in delivering a quality event service in line with the event experience. Service quality is defined by Donthu and Yoo (1998) as the difference between perceived service performance and expected service level: the difference is the gap.

This *laissez-faire* approach is not observable at landmark venues in the developed UK events industry. In a location with a highly competitive market, good reputation is the priority for attracting clients (and more pertinently, to achieving return clients) in the experience economy. Even so, in the UK as well, most event organisers are not trained in meeting consumer needs and expectations:

> Never had any training on that to be honest with you. It's not something that I've needed, I don't think. I think the industry, yes you need to have that. There's lots of personal attributes you need to have to work in a high-pressured environment dealing with lots of people pre-show, onsite, whatever. We as a company don't have training at the moment. It would be worthwhile for some people but I haven't done any of those qualifications.
>
> (UK8 – organiser for a pan-European publishing company)

This is indication that the events industry in the UK is still not closely attuned with meeting consumer expectations of service, which is a fact confirmed by Berridge (2012), who found organisers do not consistently achieve the main desired outcomes for all guests.

The above organiser feels that training in customer service expectations is *not something that I've needed, I don't think*, which reveals naivety in understanding the value of people in both the social and professional contexts of being an organiser of events. It is a 'people' profession after all, which is why entrants to the industry who study for a degree in events management are taught subjects such as customer service management and the psychology of consumer behaviours.

It is unsurprisingly a parallel situation in Poland and the UK, even though the UK industry is more developed. This shortfall in meeting customer service expectations is tolerated in both locations because in the microenvironment of an event's immediate surrounds (Reic, 2017) an event consumer will judge the experience (and service) they receive at each event to be autonomous upon reflective factors such as quality of food, how easy it was to find the venue, and even the weather on the day of the event.

But it ignores the fact that event tourists exist in the macroenvironment and will benchmark the event experience (and service) they receive against other events in the wider experience economy as they travel around in the global market for international events where political, economic, social, technological, environmental, and legal factors are less manageable (Reic, 2017). This creates a distinction between the domestic experience economy and the international experience economy, in that an organiser might learn to meet service expectations for home clients and attendees, but not when delivering to international clients and foreign attendees. Herein lies the cause of a second tier of events service and the existence of a gap.

A second tier occurs in a location where organisers evolve the learning of meeting the expectations of their home clients and consumers (clientele) and get used to delivering to that level of expectation. Vaughan (1996) refers to this pattern of normalisation during her investigation of the National Aeronautics and Space Administration (NASA) Challenger disaster (see Section 4.4 'Attitude to Risk'). The problem with the normalisation of a second tier in the domestic context is that it does not translate to higher levels of expectation across the wider market for international events. So, when international clients and foreign consumers visit an event in a second-tier destination, their expectations will not be met and they will consider that location as substandard.

The way out of a stagnant state of normalised second-tier standards is for the evolving industry to shape a construct where organisers are led to identify self-learning acculturation from the international market to know what standards are expected and how to meet them. This is possible to achieve if

organisers are making contact with foreign clients and international attendees, which is the case with all capital city landmark venues. These organisers
have opportunities to learn from observing their foreign clientele, adopting
practices and procedures that are expected by their clientele, and conducting
after-event evaluation and feedback procedures.

It is less possible to achieve acculturation learning for organisers at independent and smaller venues where they are not interacting with international
clientele. This is where formal education has value, as does organisers travelling to overseas events. But again, the industry needs to shape a construct
where organisers across the evolving industry recognise their need for a self-
learning environment.

The gap in service quality is less pronounced or even invisible in the
developed industry because organisers in this construct have evolved to meet
the expectations of their foreign clientele. The gap in service quality does still
exist, though, because not all organisers in the developed industry are working with foreign clientele. Even if they are, they might have learnt to meet the
expectations of their clientele drawn from a particular overseas location such
as, say, America or the Middle East, but not China.

Event tourists possess experience of the increasingly competitive global
tourism market. Thus, the expectations of tourists in the experience economy
are constantly rising (Augustyn and Ho, 1998). Augustyn and Ho further identify a strong relationship with consumer services and tourism and find the
SERVQUAL service quality model (Parasuraman, Zeithaml and Berry, 1988)
to be important for defining the true meaning of consumer satisfaction. The
model uses gap analysis to demonstrate how event tourism organisations can
improve their service quality:

- Gap 1: Between consumers' expectations and management's perceptions
  of consumers' expectations
- Gap 2: Between management's perceptions of consumers' expectations
  and service quality specifications
- Gap 3: Between service quality specifications and service delivery
- Gap 4: Between service delivery and external communications to
  consumers
- Gap 5: Between consumers' expectations and perceived service

There is work to be done in both the developed UK events industry and the
developing Poland events industry to educate organisers in service quality
and meeting expectations of service. It is further advanced in the UK because
the intensely competitive marketplace demands organisers to provide good
service to win and retain business and protect their good reputation. But in
both locations the customer service elements of events management are not
being addressed.

As an events industry takes shape, and when it pivots shape, organisers need to align themselves with the tourism side of events so they reach and penetrate overseas markets to attract event tourists. This can be done in alignment with a destination's place marketing initiatives. But winning foreign clients and international attendees is the beginning. Once they have been won, organisers need to meet customer service expectations which, in an international marketplace, are benchmarked elsewhere. To achieve the objective of meeting international expectations, all organisers in both the developed and developing industries – and not only those at landmark venues – should be led to recognise their need for a self-learning environment. This alone would enact acculturation learning from international clientele either through direct contact, formal education, or by engaging with professional associations for knowledge sharing.

3.6 Summary and Conclusions – Tourism and Events

Organisers in the developed events industry have created a plurality of approaches in reaching foreign markets by both immersing themselves into a foreign target market and by drawing foreign markets to their home destination. They are active in penetrating foreign markets through targeted marketing and promotion initiatives, but they also travel abroad to their foreign clients to draw bookings into their London venues. This plural of activity is due to intense competition in London with its abundance of landmark venues, which is driving organisers to need to win clients quickly by developing close relationships with bookers and agencies.

In the developing events industry, there is less demand from foreign clients, so the competitive culture is less intense. At first glance it might seem an advantage to have less competition in the marketplace, but it means organisers are not driven to fight to win business and are complacent about penetrating foreign markets.

This book, however, is not concerned with any gap between the UK and Poland in the levels of event tourism business they are winning, nor the gap in their competition culture, but the gap between their approaches to winning event tourists.

UK organisers are active and proactive in winning foreign business to capitalise on capturing the event tourism market in immersive and technological ways. This is necessary to do because European regions are increasingly competing to attract visitors with soft factors such as the location's image, which is referred to as 'place marketing' (Warnaby and Medway, 2013).

Conversely, Poland organisers adopt a non-interventionist approach and wait for foreign business to knock on their door. This is because Poland organisers do not feel the drive to be competitive in winning event tourist business, so they do not invest in knowledge, resources, or effort to attract event tourists, or reach and penetrate foreign markets, or meet the expectations of event tourist clients and consumers as a transient social demographic. This would not be a concern if Poland were a stand-alone events industry. But Poland exists within the global market for international events and is already operating in the event tourism marketplace. This means, first, that Poland organisers are missing out on business opportunities from foreign clients and international consumers; second, they have less opportunity for cross-cultural learning; third, Poland is shaping its own tier of event provision, and therein lies another gap.

For a destination to increase its visibility to foreign markets, it needs to create an identity by drawing on sociological, geographical, economic, and cultural aspects to maintain the competitive edge as an urban landscape shifting from industrial to cultural capital. Other European municipalities are becoming more professional in events policy due to their recognition that events are important for urban economy, the quality of life, and the attraction of a city. An example is Armstrong, Hobbs and Lindsay (2011) describing the place marketing undertaken to impress inspectors for the 2012 London Olympics, which included relocating undesirable elements of the public, carpeting a train station, and hiring a brass band. But event organisers in an emerging industry tend to think local and are not thinking global in their response to the globalisation of the international events marketplace to which they aspire to belong.

Tourists search for pleasurable experiences (Urry, 2002b) to escape from everyday life (Berridge, 2011) and the experience economy is about making money out of experiences. In the developed UK construct, organisers have begun to strategically align their events provision to meet the experience economy. Poland organisers, by contrast, know the experience economy exists but rely on the experience delivered by their unique landmark venue, rather than creating the service experience which international clientele expect to receive. Poland organisers also take a non-interventionist stance by waiting for the experience economy market to arrive to them by chance – drawn by the venue alone.

It requires an events industry to be actively marketing to engage with prospect consumers of both the domestic experience economy and the international experience economy who decide whether to travel to attend an event. Because event tourists are choosing to spend their money on tickets, travel, accommodation, hospitality, and merchandise, the event experience must live up to the promise and fulfil their expectations. Otherwise, the social value of the experience concept will become devalued – and global-oriented consumers are ever more demanding.

Event tourists from different locations will have different levels of expectation of the experience they receive at an event. Applying Symbolic Interaction (Berridge, 2011) will assist organisers to understand how consumers interpret experiences. But, as Berridge points out, experience is a complex series of relational components that are not uniform during the length of time it occurs or be the same for all those involved. Consumers who are from a developed events marketplace, for instance, will have wider experiences of previous events and have a wider availability of choice of which events to attend than consumers from an undeveloped or emerging events marketplace. Still, an events industry needs to get it right, regardless of their stage of development, because as soon as a country reaches a certain level of affluence the demand for services shifts to experiences (Boswijk, Thijssen and Peelen, 2007).

Evolving societies take on a new character by the constant requirement to adjust to changes induced by the evolution of a host society (de Kadt and Williams, 1976) which in the case of Poland is the European society. This means an evolving society with an industry that is emerging into the global marketplace cannot do so if the service quality of that industry does not meet international expectations. The implication for Poland is that if they, as an emerging events industry, do not shape with the socio-professional construct of the global marketplace for international events, their autonomous events industry will take shape around the limitations of its own societal environment. The impact will be to expose foreign clients and international consumers to gaps in safety, quality, and standards and a substandard tier of events within the one global marketplace for international events.

For a socio-professional culture shift to happen in the Poland events industry, international experience would be a valuable contribution to the profession as it develops the four characteristics of profession identified by Ritzer and Walczak (1986) as theoretical knowledge, self-regulated training and practice, authority over clients, and orientation to community rather than self-identity. Without the international unification of standards of international events, Poland is shaping a loose socio-professional construct with fewer rules whereby a tight culture has many rules, norms, and standards for correct behaviour, as can be seen in the UK events industry.

The events industry in both the UK and Poland are not in alignment with local and national destination marketing and place marketing initiatives. Local and national authorities and associations need to position organisers, venues, and specific events to reach event tourists as well as leisure tourists and extend inbound visitor traffic. This will enhance the profile of London and Warsaw destinations with their experiential offering to visitors through leisure events and corporate events and will attract higher levels of event-related spend at each destination such as on accommodation and visiting other attractions.

4 Risk Awareness at Events

When examining factors that shape an events industry, risk awareness must be considered because safety at events needs to be the priority of every organiser anywhere. Also, all event consumers are exposed to risks at every event, wherever the event happens to be taking place.

The particular interest of this book is impacts on safety at events if risk awareness varies in different locations due to the stage of development influencing the way an events industry takes shape.

Safety and risk management is an essential principle for events industry professionals and is key in planning, according to Bladen *et al.* (2018), who state that safety should be factored in from the start and be considered during every part of the planning process. Beech, Kaiser and Kaspar (2014) insist that risk assessment must be carried out for all events during the entire life cycle of an event so that it is an integral and strategic part of organisational processes and decision-making. The Event Management Body of Knowledge is a resource which categorises risk in the 'domains' section of the EMBOK model.

This chapter explores, first, whether event organisers in both the developed and developing constructs are risk-aware at all stages of managing events, particularly in the planning stages where the identification and assessment of potential risks can mitigate them from occurring. Second, whether there is a gap in risk awareness and the acceptance of risk due to factors within a developed socio-professional structure and one which is in development.

> *To be fair, everything should fall behind safety. Really. If you go right* [inaudible] *to the crisis management planning, everything is about the safety, wellbeing and welfare of visitors coming to site. It is the number one of everything. You have to put that one down.*
>
> (UK6 – landmark convention centre)

The approach of this organiser is indicative of professional event management and supports the statement in Berners (2017, p. 23) that 'the job of an event organiser is to eliminate or certainly minimise risk'.

DOI: 10.4324/9781032684161-4

Not all risk can be eliminated but there is a process of risk identification and assessment to mitigate potential risks (set out in Section 4.1 'Management of Risk'). This is important; as Douglas and Wildavsky (1982) acknowledge, we cannot know the risks we face now or in the future, but we must act as if we do. This helps understand the proactive or before-event approach to risk whereby risks can be managed out of an event rather than left in to cause damage or harm. Events are temporary live happenings, so if a risk is not managed out of the event before the event happens, it is already too late, and the damage or harm will have impacts and consequences which will be difficult to recover.

4.1 Management of Risk

The process of risk management is set out by Schwarz *et al.* (2017, pp. 181–184):

1 Risk identification
2 Risk analysis
3 Risk treatment
4 Risk avoidance
5 Risk acceptance
6 Reduction of the likelihood of occurrence
7 Reduction of the consequences of the occurrence
8 Transferring the risk
9 Retaining the risk
10 Risk monitoring

This ten-point process highlights the depth of risk management and it is not a cursory activity or tick-box exercise to satisfy compliance to legislation but is a thread to weave through the fabric of the life cycle of managing events. Compliance is somewhat of a get-out for some organisers who might go so far as to be compliant but go no further, which is discussed later in this chapter.

But there is social responsibility to provide a safe event, and this organiser of community events underlines the importance of risk management in their public-facing role:

> *Ensuring we're compliant with health and safety in terms of any kind of emergency procedures that I'd feel comfortable running an event, and also the public perception.*

(UK5 – organiser of local authority events)

This organiser is concerned with the social perception of their events which is the driver for them to comply with risk management procedures to ensure that the public are kept safe. To run an event with the potential to cause injury to members of the public would have a wider detrimental social impact than causing harm to an attendee at a private or exclusive event. This perspective reflects the individual versus societal risk categorisation by Gerber and Von Solms (2005), whereby individual risk is about risk to an individual, worker, member of the public, or anybody living within a defined radius from an establishment. Societal risk is about risk to society caused, for example, by the chance of a large accident causing several deaths. This viewpoint seems too definitional, however, because events bridge both categories, so fit within neither.

First, all events are for groups of people, so if there is risk to one (individual risk) there is risk to several (societal risk). Second, all events are live, so the organiser must consider, assess, and mitigate both the individual and societal risks. Third, it would be wrong for the organiser to focus their risk management process for one stakeholder group only, such as the client (individual risk) and not the guests (societal risk) or the sponsor (individual risk) and not the workers (societal risk).

Some service professionals who are employed to take decisions on risk are reluctant to contemplate risk if death is a potential consequence (Carson and Bain, 2008). It is understandable that some organisers might prefer not to identify risks which will potentially cause them a problem to mitigate, and the monetary cost of mitigating them. It might be tempting, then, to look away, pretend those risks are not there, and hope nothing goes wrong.

This type of risk avoidance is observable in the socio-professional construct of the emergent Poland events industry, but not in the developed UK construct. Even so, this is more to do with the lack of knowledge and formal education in the need for risk management processes than organisers in Poland wilfully ignoring the presence of risks. Whatever the reasons an organiser might ignore risk, risk must not ever be ignored.

Organisers in the UK events industry are further ahead than Poland in their understanding of the need to mitigate risks and the value of doing so. Bladen *et al.* (2018) note that the UK's perception of health, safety, and risk in the past tended to be red-tape bureaucracy that ticked regulatory boxes while creating extra work for the event organiser. This may have been so, but it did have the effect to shape the UK events industry towards an audit culture psychological stance in line with other sectors. It also created the socio-professional environment of safer events. It is easy to view red-tape bureaucracy as a negative factor, but it has in fact driven organisers in the UK to compliance behaviours because of the fear of legal repercussions and litigation if something were to go wrong (negligence).

Thus, organisers in the developed industry are wont to be compliant to avoid litigation, which means they must be educated in the regulations and legislations, or risk breaching them. An individual organiser or the company they work for might recognise this need and seek training, accreditation, or knowledge from a professional association or specialist. It is therefore the individual organiser or the company they work for which has driven the UK events industry to professionalisation and risk aversity rather than consumers or even the insurance industry responding to previous tragedies at events.

This contrasts with the emerging industry where the fear of repercussions and litigation is not as pronounced and organisers are further distant from the legal system because it is less likely for an event to receive a visit by an authoritative body. It means Poland organisers operate in a culture where they feel they can get away with noncompliance – not because they wilfully aim to breach safety legislation but because they do not feel the need to bother to think about it in great depth (or pay for mitigative steps). The shape of the industry in Poland, therefore, is different from the UK because there is less drive by the law to professionalism if individual organisers or the company they work for is not rushing to be compliant because they have less fear of litigation.

4.2 Learning about Risk

When it comes to risk, an emerging events industry can learn by observing what has happened elsewhere and then adopting corrective measures before having to go through the learning from their own painful experiences. Recognising how a developed events industry has shaped itself towards an audit culture to prevent disasters from happening would speed the professionalisation of the emergent industry, driven by the need to be compliant.

The need for health, safety, and risk procedures has been accentuated by milestone event disasters that could have been prevented if correct health, safety, and risk procedures had been in place (Bladen *et al.*, 2018). There is an evident milestone with the bombing of the Manchester Arena Ariana Grande concert on 22 May 2017 where 22 concertgoers were killed (Gardham, 2020). The first volume of the government Inquiry into the bombing, published in June 2021, set out to identify missed opportunities and failings of the accountable parties being the venue operators, the contracted security company, and the police (Saunders, 2021). The Inquiry chairman stresses that the purpose of the Report is not to apportion blame but to learn what went wrong and reduce the risk of such an event happening again and, if it does, mitigate the harm it causes. Bladen *et al.* (2018) say that such occurrences result in governments and industry associations regulating against future problems by increasing legislation, thus reshaping the industry.

One example of a law update stems directly from the Manchester Arena bombing. Martyn's Law (Gov.uk, 2022) mandates, for the first time, those held responsible for premises and events to consider the risk of a terrorist attack and how they would respond. This law states 'simple steps save lives' to ensure venues are better prepared and ready to respond to a terrorist attack.

Increased regulation, such as Martyn's Law, and others, becomes a resource for event organisers to comply with in ensuring their events are safe.

This is the point where an emerging events industry can shape itself to fit with the behaviours of organisers in an advanced events industry to close gaps in risk acceptance and risk avoidance.

This viewpoint is underpinned by a UK organiser who is a member of the International Compliance Association (ICA), which declares itself to 'provide resources that demonstrate commitment to the highest standards of practice and conduct, enhance professional reputation and employability, and enhance the performance of the organisation' (*ICA*, no date):

> *We offer emergency procedure briefings, also under the CDM* regulations that came in 2015 we now need to do site inductions which is driven by the organiser and the responsibility of the organiser that we will make our staff go through it.*

> (UK6 – landmark convention centre)

This organiser understands their professional responsibility in society as an organiser to lead their staff to deliver a safe event. However, in the Poland's evolving events industry the emphasis for an organiser is on the delivery of an event rather than risk management for the delivery of a safe event.

Professional bodies in the UK, such as the CDM, provide organisers with a socio-professional support network indicative of a functioning 'art world' (Becker, 1984). This has driven the events industry forward, with people supporting other people to achieve the art product – in this context, the event. It is the art world which influences the way organisers learn to do their job and reflects the Douglas Grid Group Diagram (Anfara and Mertz, 2011), arguing that people need to fit with the behaviour of the group to which they belong.

Participation with professional associations is a choice for organisers in the UK developed industry – there are many to choose from, which is counterproductive because it is confusing, and an organiser might find it simpler to engage with none. In the evolving market in Poland, there are conversely too few associations to choose. This is a gap between a developed events industry and one that is evolving, but not a significant gap because not all UK organisers participate with industry associations, and those whom do can choose which ones and how many. Because there is no legal requirement to belong to one or any professional association, the gap exists in the UK industry anyhow.

In Poland it is the same gap, albeit more widespread. This gap is not due to the stage of development of the events industry in each country. This is what makes it an insignificant gap between the UK and Poland and will only get filled if both countries require organisers to join a professional association for accreditation.

---

*   The Construction (Design and Management) Regulations 2015 (*CDM*, 2015).

This said, there is a need for organisers in the UK to comply with constantly evolving legislation brought about by milestone disasters. This is more pronounced for high-level organisers at landmark venues, and less driven for organisers at smaller, independent venues. Nevertheless, there is an appetite among organisers in the UK for knowledge sharing of best practice:

> *The typical ones for us would be people like the Association of Event Organisers and the Association of Event Venues . . . sharing best practice and I guess really helping to elevate the standard of our industry, if you like. Because if we can do it better. . . . There are two elements, for us we'd be in competition with other venues but actually the other side of it as a group of venues across the country it's our job to collectively improve our standards and offer our customers the best . . . It's about trying to elevate our own industry. . . . We need to be presenting the united face of the industry.*

(UK4 – landmark convention centre)

This view reflects the social awareness of organisers in their collective responsibility to advance the UK events industry. Knowledge is the objective here, which Brown and Stokes (2021) argue is at the centre of events management. This provides a snapshot of the shape of the UK events industry and reveals a stark differential between the UK and Poland because organisers in the UK are keen to exploit their network for knowledge-sharing activities and associations. Poland organisers are autonomous and will view other organisers as competitors in the marketplace. It is a professions difference because organisers in the UK have been driven to develop a sociocultural behaviour to be professionals and to professionalise their habitus:

> *I have a *NEBOSH qualification and **PRINCE2 project management qualification.*

(UK6 – landmark convention centre)

This organiser is proud of their NEBOSH and PRINCE2 accreditations. For them, possessing risk management qualifications provides skills and competences accredited by the industry, which reflects the shift in discourse from the 'sociology of professions' to the 'sociology of professional knowledge', and

---

* NEBOSH (National Examination Board in Occupational Safety and Health) is a recognised professional examination in health and safety (*NEBOSH*, no date).

** PRINCE2 (**PR**ojects **IN** Controlled Environments) (Pielichaty *et al.*, 2017) courses begin from £540 (*PRINCE2*, 2020).

then to the 'sociology of expertise' (Macdonald, 1999; Eyal and Pok, no date; Young and Muller, 2014).

Knowledge gain is important not only for this organiser to do their job (because they could do the job without qualifications, as others are) but because they feel professional among their professional counterparts in a professional setting. This organiser can, for example, assert authority over others in the industry construct, including clients, because the organiser possesses knowledge and confidence in an area of event specialism which others might lack. This allows an organiser to elevate their status and is an example of 'the real event manager' (see Berners, 2017, pp. 19–21). It confirms, then, how an individual organiser or the company they work for is driven to professionalisation by recognising their need to be compliant.

Possessing industry qualifications and knowledge will make headway in the production of safer events wherever risk management is being practised. This is a good thing, of course, but it does drive the shaped events industry towards accusations of credentialism. Brown (2001) explains this can lead to declines in employers' use of other recruitment criteria by insisting on candidates who have a certain set of professional qualifications or an events management degree, say, thus homogenising the social landscape.

At least in the shaping industry there are no such barriers of entry because qualifications are not required. For all the deficiencies that the lack of professional qualifications may entail, it is certainly not a credentialism social landscape which limits diversity in the sector, unlike which has taken shape in the UK events industry.

Due to the different stage of development between the events industry in the UK and Poland, courses in events management are not yet available in Poland and so are not recognised. Therefore, qualifications in events specialisms are valued in the UK but have no value in the Poland events sector:

> *Huh* [laughs], *it's quite a long story, I used to work in different kind of companies because I worked for 15 years in the mobile telecommunications sector. I did some marketing there. That's why I'm here at* [venue name] *to somehow support marketing and to organise events. But most of my career was in a different place and different sector. I was fed up of* [inaudible] *world – rat race and that kind of stuff. . . . I do not have any theoretical background but rather I would say some practical one.*
>
> (PL2 – landmark venue)

By laughing at the outset of the interview, this organiser may have found it incredulous to be interviewed as an event specialist because their professional background is in other sectors. Nonetheless, this is an organiser at a landmark venue in a European capital city (Warsaw). This organiser is not an imposter and might be surprised to know that their situation is not dissimilar to London organisers who do not have a theoretical background in the management of events. This is the reason both the evolved and evolving industries place value in experience over qualifications and why the events industry in both the UK and Poland mirror the shape of holding experience above qualifications.

This is at odds with the theoretical works of Ulrich Beck and Anthony Giddens that we are living in a developed risk society (Ekberg, 2007). Their theory is that there has been a reconfiguration in the traditional way risk is identified, evaluated, communicated, and governed. It is not only the sum of probability of an adverse happening and the severity of the consequences, but now it includes the perception of risk, the communication of risk, and the social experience of living in a risk environment. This, they argue, is the way to effectively manage risks emerging in the age of modernity of the risk society. Beck (no date, cited in Boyne, 2003) considers that there is a break within modernity in shifting from classical industrial society to a new self-endangering civilisation. Perhaps, then, event organisers are focusing reliance on their experience to redress their lack of professional qualifications in meeting the modern risk environment:

> *We had to learn a lot by ourselves. Everything. We gained experience whilst organising things because we always had to deal with these problems, and we always had to see to it that everything was tip-top. That there would be no problems, you know, during the event. We had firefighters, we had security, we had medical, we had special guards from the* [name of venue] *as well.*
>
> (PL3 – organiser of charity events)

This interviewee – being an event organiser in the emerging Poland construct – has no access to formal education and so has to learn from their experiences, and is responsible for events attended by up to 5,000 people. Still, their reliance on trained *professionals* such as *firefighters, security, medical*, and *special guards* is no different to the level of specialist support required by qualified organisers. As previously identified, an organiser having to learn from their experiences means they will make mistakes before learning from the pain. In many cases their mistakes cannot be rectified because the event is happening and will soon be gone. In some cases, their mistakes might lead to harm, loss, or reputational damage. They could, as already mentioned, learn from other organisers' previous mistakes to avoid them being replicated.

4.3 Acceptance of Risk

The level of acceptance of risk is not impacted by specialist qualifications or education in events management because some organisers with specialist education do not enact risk management procedures such as setting objectives and conducting after-event evaluation. Specialist education does, however, inform organisers how to enact risk management procedures, which leads to knowledge of risk procedures, risk control, and risk management being fundamental in managing events.

An emerging events industry can learn about risks at events (and shape their construct) from tragedies that have happened in an already-shaped industry so as not to learn from the painful experiences of its own tragedies. For this to happen, it will require international policy transfer and imported knowledge from graduates of events management whom have been educated by studying past tragedies, risk milestones, and the legislations and practices which have shaped an events industry after it has experienced tragedy. The barrier here is the lack of risk reflection in event literature of risk milestones that have shaped a developed industry. Mostly, event literature focuses on potential risks, not past risks.

Another problem here is the gap between rules and what people do in actual behaviours. Triandis (no date, cited in Gillespie and Hennessey, 2016, p. 112) views this as dependent upon the construct of either a 'tight' or 'loose' culture, whereby a tight culture has many rules, norms, and standards for correct behaviour as can be seen in the UK events industry. In the UK, rule breakers risk receiving criticism or punishment or at least judgement and damage to reputation. This concept has relevance with events because of their temporary nature making them less likely to be identified as noncompliant before they are gone, and the organiser less likely to be criticised or punished which drives towards a loose culture, observable in Poland.

The basis of social order is conforming to social rules and norms, and deviance occurs when norms are broken (O'Donnell, 1997). Being less likely to get noticed, criticised, or punished encourages organisers in Poland to be less driven to professionalism and to perform deviant behaviour. This is also visible in the high level of risk acceptance of other emerging event constructs such as Russia hosting the Sochi Winter Olympics with 'hundreds' of workers losing their life (Shaw, Anin and Vdovii, 2015) and Qatar with the World Cup and 6,500 deaths (*The Guardian*, no date).

Partial compliance behaviours of organisers can be observed at events in Poland, where rules and norms do get broken. 'Partial compliance' is where organisers have identified a hazard and go partway to mitigating the risk of it causing harm, but without fully negating that potential consequence from occurring. By conducting partial compliance, an organiser might not be accused of ignoring the hazard because they have identified it, first, and have gone some way in mitigating it, second – albeit not going so far as to fully mitigate the risk from potentially causing harm.

That last bit of not fully mitigating the risk is deemed to be 'worth the risk' because it is unlikely that it will be noticed before the event is gone. Anyway, for it to be noticed would require a safety official to visit the venue, which is less likely in an underdeveloped events industry.

Another reason for ignoring risk, or at least taking only partial steps, is the notion of fate, chance, or having faith in God to protect people. In the Poland construct, this is a real concept due to the devout Catholic society. Organisers might, then, possess an embedded belief that they can trust in chance and fate because God will protect.

Acceptance of risk at events by organisers in the developed industry is lower than in the developing industry because it is more likely to be noticed in the former, and if it is, it could halt the event or incur litigation. Silvers (2008) attributes this to the variance in different cultures and different locations, and Boyne (2003) concurs that standards of acceptability may vary across cultures.

Operating within a loose culture, the following organiser recognises variances in risk acceptance behaviours at other events and is able to determine what is good practice and what is not, from experience only:

Sometimes I do come across things which for us would be unthinkable; poorly managed; poorly organised.

(PL4 – organiser of charity events)

This organiser feels able to differentiate their own events from others and is benchmarking good practices against observing lesser practices. This has positive implications that the role of tragedy facilitates organisers in an emerging industry to learn from the tragedies which have shaped an industry that is further developed.

4.4 Attitude to Risk

It's about health and safety. So, what to do if something goes wrong, if somebody is taken ill and how you can get help quickly. It's about who we might be expecting, who we need to look out for, any VIP guests. It's things like, you know, security. Security can be a big one for some events. About looking out for odd behaviour or areas that people should not be going into and how to raise help if we need it. Just making sure everything is as we would want it to be.

(UK2 – organiser at an institution)

The use of future tense in the above statement suggests that this organiser considers risk during the planning stage of their events to identify and anticipate what those risks might be. This is underpinned in the statement, *what to do if something goes wrong*, which indicates a forward-thinking stance and not waiting for something to go wrong before thinking what to do about it.

The use of the phrase *who we might be expecting, who we need to look out for* further corroborates their forward approach to risk by knowing in advance who will be attending and looking out for them to arrive.

Even though this organiser does not routinely set objectives (to have a safe event is an objective) or conduct after-event evaluation (to measure that the event was in fact safe, and to identify what risks could be mitigated for future events), they are attuned to the presence of risk, which makes their behaviour risk-averse.

Haniff and Salama (2016) recount three levels of risk attitude:

1   Risk-averse – risk is avoided and opts towards less risky events.
2   Risk-seeker – the higher the risk, the higher the returns.
3   Risk-neutral – decisions are based on detailed analysis.

Of the above, all events must start from the perspective of risk aversion where risk is avoided. But there are examples where organisers delegate their responsibility of risk management:

> We would have a supplier's meeting pre-event three or four weeks before an event. I would present how we saw the whole event running. When we go onsite at venues they will do sort-of fire safety briefings. . . . They will only do a fire and safety briefing of the venue, of where to go.
>
> (UK8 – organiser for a pan-European publishing company)

The above statement reveals that the organiser has an expectation that suppliers will take responsibility for *sort-of fire safety briefings*. This is understandable because delegation to other professionals is acceptable, had the terms *sort-of* and *only do a fire and safety briefing of the venue, of where to go* had not been used because these ambiguities indicate a level of complacency on top of delegating the utmost priority to a supplier.

This type of risk-deferment behaviour reinforces that some organisers accept that there is the presence of risks at their events but are not prepared to accept responsibility for mitigating them.

The question to this organiser, and others who delegate the responsibility for risk, is what would happen if something went wrong at their event? It is likely that the blame would be directed towards the supplier who was charged with delivering the fire safety briefing. This would be evidence of a blame shift.

Douglas (2003) discusses risk and blame, where she cites Hegel and Weber, identifying that knowledge and awareness increases technical control.

If organisers delegate the responsibility of risk management to other parties, they will fail to evolve their own knowledge and awareness of risks that are present at their events and plan the measures to control them.

Pearn, Mulrooney and Payne (1998) discuss turning blame cultures into gain cultures by learning from mistakes, which all organisers should be doing. But if an organiser is not involving themselves with the risk aspect of their events because they delegate that responsibility, they will not gain but they will blame. This amounts to lost opportunities for organisers to learn from experience, if experience is not gained.

Learning from what happened is the objective of the Inquiry into the bombing of the Manchester Arena – seeking to answer the question of accountability without apportioning blame but to identify who was accountable for the missed opportunities to prevent the bombing or mitigate the harm it caused. An incident of this magnitude will reshape even a developed events industry and will update legislations and practices. There are even wider implications of opportunity, other than legal, for organisers outside the UK to learn from such tragedy. But this will happen only if there is international and cross-cultural engagement in place to facilitate acculturated learning.

It is important for organisers to work with risk so they learn to identify, assess, and mitigate risk from causing harm. Cottle (no date) writes that contemporary risks are historically unprecedented, so their potential catastrophic effects are invisible. He states they can only become visible when defined within knowledge, the legal system, and mass media. He further asserts that this is a social-constructionist formulation where risks are dependent upon how they are made socially visible.

Until a tragedy occurs, then, the issue with risk acceptance is normalisation, which Vaughan (1996) concluded with her investigation into the NASA Challenger disaster. In determining how Challenger was launched against objections from engineers with a design known to be flawed, Vaughan revealed a culture of incremental descent into poor judgement and a pattern of normalisation in engineering risk assessments. There is danger of this attitude happening with events in a shaping industry if the societal acceptance of risk becomes the embedded culture of that location. This is very likely because nobody in that social construct knows any better. If this happens it will shape a localised tier of lower risk aversion compared to known and accepted levels of higher risk aversion across the global marketplace for international events.

The question is, how will an organiser in Poland contrast the practices in their own industry, with which they are familiar, against those in an industry that has already shaped from its tragedies, and identify the disparities?

The impact of this gap is that event tourist clients who book their events into a location with a shaping events industry, and event tourist consumers who travel to attend an event in a shaping events industry, will be unaware of their exposure to a higher level of risk acceptance and lower level of mitigation of risks. This disparity already exists:

> *International clients almost always would sit with us and expect a debrief to see what can we learn, how the event went. What can we learn from it, what could we do better, what did work, what didn't work. Sit and think about how did it go. Very important to see if the client is happy and if their goals were met.*
>
> (PL7 – organiser of events)

This organiser identifies that a difference exists in the approach of international clients expecting an after-event debrief, whereas their Poland clients do not have that same expectation. It confirms approaches vary from location to location which is why the gap of risk acceptance needs to be at a consistent low level across the global marketplace for international events.

Wynn-Moylan (2018) asserts that structured evaluation after an event should lead to improvements in planning the next event, changes needed to the venue, current or potential problems, ways of improving the system, and feedback from staff. The organiser above clearly recognises the value in learning from their international clients: *What can we learn from it, what could we do better, what did work, what didn't work.* Yet they still appear to accept that their Poland clients do things differently and there is no debrief with these clients to learn about how the event went, how the venue performed, and what could be done better such as risk identification and risk mitigation.

Another organiser confirms the existence of this same gap:

> *There is a difference between Polish – maybe ten years ago in Poland – and agencies from Europe, mostly Western Europe. Because the Western European agencies are very used to having feedback so they are sending letters with 'maybe you can consider something'. Usually, they are pleased, I must say, with our work. But for the Polish companies they say 'thank you, okay, bye. Now it has slightly changed.*
>
> (PL5 – landmark historic venue)

Again, this organiser is being offered opportunities to improve the service they provide to international clients because this category of client is giving feedback on where to improve. But their Poland clients do not give feedback. Again, there is recognition of the importance of feedback from international clients, yet it still has not prompted this organiser to acquire feedback from their Polish clientele.

However, their use of the phrase, *now it has slightly changed*, suggests international knowledge transfer is happening and will help to reshape the

Poland events industry towards international practices and narrow this gap, which is reinforced here:

> *I pick up so many things, even in Vienna, you wouldn't believe it. At the Polytechnic* [venue in Warsaw] *for example I made a point that guests do not see caterers with their stuff. I said, 'you cannot have this, so think of something'. The tables on one side of the hall and they had blinds brought down and it was cut off for the public so they couldn't go there and see everything. I told myself that also the sound engineer must cover his stand with special material so you don't see it – he's not there, if you know what I mean; everybody has to have black suits which was good, and also the technicians. Lots of stuff, lots of stuff, lots of stuff.*
>
> (PL4 – organiser of charity events)

This organiser demonstrates their willingness to learn from sources not available in the Poland events industry construct and appears excited and pleased for picking up on these elements and importing the learning to their own events. Although this organiser is referring to aesthetics, it is reasonable to expect that their acculturated learning will extend to other areas of managing their events, such as risk management, learning from tragedies elsewhere, and adopting a lower threshold of risk acceptance.

The problem is that organisers in a developing industry receive less experience, unless they are involved with events in a foreign location with perhaps a further evolved socio-professional construct. If they are involved with events for foreign clients and international attendees, they can gain experiences and learning, plus learn from the experiences of tragedies, which have already happened elsewhere, without having to learn from the painful experiences of their own failures – so long as they recognise these learning opportunities and are willing to adopt the learning into their practices.

4.5 Learning From Experience

People learn from the mistakes they make while doing the job. This is what builds career experience and is why industry experience is highly valued by employers on a candidate's curriculum vitae. In many cases it is why some firms and individuals are trusted implicitly and their word and abilities are considered more significant than any qualification – even in the recruitment process where qualifications should matter.

However, where organisers rely on their career experience to learn from, rather than a formal education, it takes much time and can be too late for mitigating risks at events. Events are consumed at the point of delivery, plus they cannot rehappen, so if a risk is left to cause a consequence it cannot be recovered. This is more prevalent in the developing construct because organisers will not yet have access to formal education in the specialisms of managing events. Therein lies the cause of a gap, even though the UK events industry eschewed formal education or qualifications until recently.

In addition, it would be helpful (and safer) to learn from the mistakes which other organisers have already made than to learn from one's own mistakes as they happen and those which are yet to happen. This is imperative not only for achieving client and consumer satisfaction, quality, and meeting the event objectives, but it also becomes real and essential with mistakes that can be determined as 'tragedy'. Examples are the Manchester Arena bombing in the UK construct and the murder of the Mayor of Gdansk onstage at a festival in the Poland construct (Buras, 2019).

Events literature does widely provide insight with risks at events – some event textbooks are devoted to the subject – but it tends to be practical operational guidance rather than cross-cultural studies or milestones of tragedies, legislations, and practices from which an organiser – or student – might learn. For example, Wynn-Moylan (2018) proposes an Event Safety Management System, which is a set of planning activities leading to a formalised set of procedures for the management of safety at events and venues. He asserts that an event safety policy defines the commitment to society that the event organiser makes in hosting a safe event. So, the delivery of a safe event is not an option but a societal expectation. Yet learning from the history of tragedies at events is widely ignored and the milestones which have shaped a developed industry are largely uncharted.

There are other procedural approaches to risk including the EMP, which is a document that gets disseminated to a Safety Advisory Group (SAG):

For any event, you prepare an EMP, which is an Event Management Plan. That EMP is usually produced twelve weeks before an event at a minimum. That will be presented at something called a SAG Group, Safety Advisory Group, and then you have feedback from the statutory authorities around that table.

(UK5 – organiser of local authority events)

This organiser stating *for any event* indicates the EMP is a procedure for all their events to facilitate feedback from the SAG made up of statutory authorities. This means that an organiser does not have to be a risk or safety specialist but can gather a group of advisers with authoritative knowledge, experience, training, and qualifications to support the event, provide advice, and ensure that it is going to be safe. This is important for organisers operating in a developing events industry to know because it facilitates them gaining specialist knowledge from expert sources, rather than their own formal education or years of experience.

A detailed account of the function of the SAG at the 2012 London Olympics is presented by Armstrong, Giulianotti and Hobbs (2017, pp. 133–141), which they refer to as the Olympics Command Zone. The assembled SAG is flexible and adaptable by necessity for each situation at each event.

Paraskevas (2006) sees this as a living co-evolving system for effective response to crises.

Working with an SAG is a learning opportunity for an organiser. However, not all events warrant an SAG because the event might be low risk, basic, and routine. In the UK, just as anywhere else, there is no guarantee that an organiser will learn from working with SAG specialists. This leads to the way most organisers learn about risk: from their own experience, which obscures the potential (and need) for learning from other organiser's tragedies that have already happened elsewhere, and the role of tragedy in shaping an events industry. This obscureness also masks an organiser's recognition for the importance of creating for themselves an international cross-cultural learning environment. What needs to be intrinsic to all organisers, then, is the recognition to formalise their learning.

4.6 The Self-Learning Environment

A lot of our events are repeat annual events so just because it's finished doesn't mean it's not immediately starting again for preparation for the next year. What we are very, very keen on at the moment is understanding where we can improve after an event, as soon as possible. We task our event managers now with starting the debriefing process six weeks before the event even comes to site. . . . I presented at the AEO [Association of Event Organisers] *three weeks ago at their ops forums and I talked about debriefing an awful lot and saying how it needs to loop continuously so it's almost not a debrief but an ongoing process, a continuous event life cycle – just because it finishes on* [date] *doesn't mean the event has finished, it's actually going to start again for next year so your planning starts the moment it finishes. And we look at the way this continuous event life cycle loops and how it feeds, and how we can continuously improve.*

(UK6 – landmark convention centre)

This organiser understands the cyclical need of learning from one event to the next as a formalised and structured process. This is not carried out alone, but in cooperation with *venue and organiser and contractor* – what this organiser calls the *three pillars to making a successful event*. The difference here, from learning from an SAG, is that this organiser is steering their own learning by knowing to engage with people in their habitus. Working with an SAG provides the opportunity for passive learning, but in the frequent and most-often absence of an SAG, an organiser can conduct active self-learning.

In the above excerpt, the organiser says, *I presented at the AEO*, so they have in fact elevated their status from learner to teacher. And the way this organiser describes the continuous event life cycle as a *loop* and an *ongoing process* reflects their recognition of the need to implement continuous evaluation for their venue to become a learning organisation, achieve their goals, and meet standards (Getz, 2018).

Allen, Alston and DeKerchove (2019) agree that employees must perform safely and in compliance with requirements, as well as continuously seek ways to improve operations and processes. Individual organisers, therefore, need to shape their socio-professional construct to facilitate a self-learning environment – the industry will not lead it. This is especially important for organisers to understand in locations where there is limited or no access to formal education, and organisers are less likely to learn from experience because they do not conduct structured evaluation and feedback procedures:

> *If everything is okay, we have another event next day or we have some days with four events so there is no time for it* [evaluation]. *So, if everything is okay, we just need to do our paperwork after this and rub our hands and forget it.*
>
> (PL5 – landmark historic venue)

The previous organiser was aware of creating opportunities for self-learning, whereas the above organiser says they *rub our hands and forget it*. Getz (2018) highlights the need to measure returns: Return on Investment (ROI), Return on Objectives (ROO), and Return on Experience (ROE), but some organisers omit these measures by saying *there is no time for it*. Where organisers fail to enact procedures from which they could learn, they are inclined to blame lack of time (see also Section 2.1.5 'Event Objectives'). This is true of organisers in both the UK and Poland, as if they know they are wrong to be omitting some procedures and are reluctant to admit that it is their fault:

> *There is never time to debrief customers. Customer is always running away and* [laughs] *chasing other subjects and we actually marching into the next project, so it is very difficult. But we always try to debrief on email or conversation on the phone maybe to catch up later on. But I know international clients almost always would sit with us and expect a debrief.*
>
> (PL7 – organiser of events)

This is another example of an organiser blaming time for failing to conduct after-event feedback and evaluation procedures for their self-learning from one event to the next. The post-event procedures are even more important where there is no access to formal education in events management. But they are not getting omitted because of an unwillingness to learn (every organiser wants to do a good job and run a safe event), it is the lack of recognition of the need to create a self-learning environment.

Organisers in the developed industry do have opportunities to learn from formal education and professional associations, so this is a knowledge gap between the developed and developing industries. All organisers everywhere can conduct active self-learning, though – it just takes them to recognise that need. An events industry needs to form this behavioural understanding as the shape of their socio-professional construct, especially where the industry is eschewing any formal education or qualifications in events management. Where this eschewing is happening, there is tension between experience and academic qualifications because experience is valued over academic qualifications in the belief that if qualifications are not available, they cannot be needed.

4.7 Consumer Risk Perspective

Risk and uncertainty in events are not limited to how an event is planned and executed but extend to the organiser's responsibility for protecting consumers attending an event.

Also, the risky behaviour of consumers is a factor at events and often exacerbated by large crowds and the presence of alcohol and drugs. Such factors are common to events all over the world, so the potential to share knowledge across an international platform from a shaped industry to one which is shaping, and facilitate learning from previous tragedies, is very relevant.

Also, the trend for providing experiential events for fun, adventure, and thrill means consumers are closer to the risk of injury and harm while being carefree with their personal safety which they place into the hands of an event organiser. Bladen *et al.* (2018) state that event design is among the greatest challenges facing the events specialist due to clients demanding ever more creative events for their attendees to experience. Clearly, in the face of these trends, it must be beneficial, and safer, for an international approach to learning from tragedy and to shape emerging and developing events industries from those which are already developed.

Consumers have a societal subliminal expectation to be safe at an event and there is a distinction provided by Douglas and Wildavsky (1982) between voluntary and involuntary acceptance of risk by consumers. They write that if people wish to expose themselves to risk, that is their own business and voluntary, but if they are unaware of the risks or associated risks it is involuntary. Their answer to this is 'better information' so that people either reject a known risk or seek additional compensation for assuming it. This means involuntary risk is due to ignorance but can be converted to voluntary risk by knowledge. Organisers should understand this when producing events and marketing them to consumers.

The consumer risk perspective needs to be considered by organisers by asking themselves whether the risks are being communicated to the consumer, so that the consumer can voluntarily expose themselves to those risks. This is especially important with gaps dependent upon location and attendees being unaware of this fact (involuntary acceptance). However, this is not what is happening because instead of considering the consumer risk perspective, organisers tend to focus on the commercial objectives:

> *With an event, obviously there's usually key drivers. So, it's either a finan-cial, or a community, or best-practice initiatives, really. If there's a finan-cial driver to an event we'd want to know all the information upfront in terms of what the costs were, what the percentage income, what the break-even was, obviously what the community benefit was – engaging with the residents and engaging with the local community. The economic benefit as well, in terms of gauging how many visitors we maybe could get to an event, what the secondary spend opportunities were around that . . . but I think you need to be very clear on what your aims and objectives are so you can actually achieve them, otherwise you could end up delivering something which isn't quite as good as you wanted it to be and it doesn't achieve the objective what you set out with.*
>
> (UK5 – organiser of local authority events)

This organiser is concerned with financial targets and commercial objectives because they organise events on behalf of a local authority and is naturally driven by budgets, spend, and Return on Investment (ROI). But local author-ity community events are organised for people – as all events are – yet this organiser is leaning closer towards financials than consumers. Additionally, this organiser of community events is exposed to societal risk foremost and individual risk secondly, so the effect of something going wrong is greater (see also Section 4.1 'Management of Risk').

Bowdin *et al.* (2011) remind organisers that they have a legal responsibil-ity to provide a safe event, and Wynn-Moylan (2018) points out that Duty of Care is a legal principle which places the event organiser as the responsible person to avoid acts of omissions that could injure another. As the legal impli-cations become more impactful when an events industry matures (because legislation matures with the industry), organisers indeed welcome and are willing to accept and engage with an international approach to the role of tragedy in shaping their industry.

As it stands, however, none of the senior-level organisers at landmark ven-ues who were interviewed included 'keeping people safe' as an objective of their events – even though risk can be defined as the likelihood of an event not fulfilling its objectives (Allen, 2009).

Commercial objectives are certainly easier to define than consumer safety because statistics (costs, revenue, profit) can be measured quantitatively. It takes greater effort to obtain, evaluate, and measure qualitative objectives. The stubborn issue is that organisers put on an event for financial reward first, not for the safety priority – which is why they will focus on their financial objectives at the cost of the risk objective:

Our objective is to get the people in and out the building, get them to empty their pockets – being very commercially driven [laughs] *– and leave with a smile on their face. I use my cynical commercial outlook but we're there to service people and keep people coming back, that's the nature of what we're trying to do. We get around 1.5 million visitors a year. . . . We've tried things like customer thermometers which actually work – smiley faces. There are some good tools out there and I would encourage more venues to engage with those tools and really benefit from them.*

(UK10 – landmark exhibition centre)

This organiser is at a significant landmark London venue which has a good reputation to protect, and they are not dismissing the value of their consumers – this is clear in the above excerpt referring to *customer thermometers . . . smiley faces.* But even when talking about consumers, the perspective of this organiser is to view people as a commodity to reach the financial objective by getting *them to empty their pockets.* Thus, this organiser is interested in the consumer perspective only because of the need to *service people and keep people coming back* for them to continue spending money and keep the venue commercially viable.

This is not out of step with every organiser who was interviewed. So, although this may be a troubling and valuable finding, it is not a gap between the UK and Poland because organisers in both constructs failed to mention consumer risks and safety. This does indicate a shortfall in the consumer risk perspective overall, rather than any difference caused by the stage of development of an events industry.

There is also stubborn focus on financial objectives in existing theoretical event literature, including omitting safety as an event objective. Examples include Shone and Parry (2013, pp. 139–140; p. 333) discussing objectives and financial planning within several sections of their textbook under the headings of 'cultural events; financial; leisure events; new events; organisational event; personal event; planning; setting', but objectives is not included in the 'safety' section. Their book does discuss risk, but risk and objectives are without linkage. Silvers (2008) only arrives at the consumer risk perspective in the very last chapter of her *Risk Management for Meetings and Events* textbook, although Wynn-Moylan (2018) does devote three chapters to the consumer risk perspective in his *Risk and Hazard Management for Festivals and Events* textbook.

Overall, there remains a gap in the literature in addressing consumer risk and safety to impart the awareness to organisers – particularly students of events management – that consumer safety is the priority objective ahead of other objectives such as financial goals and targets. It should be remembered that if an event turns out not to be safe, it can compromise the financial objectives anyhow – customers can complain or demad refunds which impacts profit and return business. To shift the discourse of this perspective would have a behavioural impact in the UK events profession because students on formal courses will import their learning to the industry. It would not yet impact the Poland events industry where there are no courses in events management. Although, if an international construct were to happen, Poland would learn from the UK and others.

Organisers in both the developed and developing events industries are aware of risk, but there is higher likelihood of repercussions and litigation in the UK if noncompliance is identified or something goes wrong at an event. This has driven UK organisers to recognise their need to be compliant with regulations and legislations. In turn, this has shaped the UK events industry to a risk-averse socio-professional construct. In Poland it is 'worth the risk' to accept risk because it will likely not be discovered. This difference reveals that compliance with the law is the driver for individuals or the company they work for to be professional, rather than consumers or the insurance industry responding to previous tragedies at events.

Organisers everywhere would benefit by learning from academic qualifications, but the industry places trust in firms and individuals' word and abilities from their experience, which is demonstrable by reputation, client portfolio, testimonials, and track record of delivering successful events. The events industry does not lead organisers to achieve academic qualifications and to a large degree eschews any formal education or qualifications in events management. The socio-professional structure takes form around organisers organically evolving their learning unaided and relies on individuals or the company they work for to identify learning needs by themselves, which tend to be driven by the need to be compliant.

Thus, an industry taking shape recognises experience as more valuable than qualifications in the belief that if qualifications are not available, they cannot be needed to run an event. But, as an industry matures and academic qualifications become available, entrants to the industry seek to study to make themselves more employable, and employers respond by advertising job vacancies requiring candidates to be qualified. It should be noted that this is still organic development by organisers and is not being led by the industry, so there are naturally-occurring tensions between academic qualifications and experience.

Such tensions are due to experienced organisers abstaining from gaining qualifications which have now become available, an understanding in the industry of the 'right' level of experience versus the 'right' level of qualifications, and the social barriers of organisers thinking they know how to do the job, so they devalue qualifications and qualified organisers, and their rejection of academically qualified entrants who lack experience.

This is where an international cross-cultural construct for knowledge sharing and learning will facilitate an emerging industry to be safer in its practices of managing events by learning from tragedies that have already happened elsewhere. Also, if organisers do not routinely set objectives, omit the risk management procedures, delegate the risk responsibilities, and fail to conduct after-event evaluation, they will not identify or learn from their own experiences of mistakes.

Therefore, the most significant factor for an emerging events industry in addressing risk is for the socio-professional construct to shape an active self-learning environment where organisers professionalise by identifying learning opportunities from other organisers, specialists, and outside sources such as their foreign clients and international attendees.

4.8 Summary and Conclusions – Risk Awareness at Events

This book is interested in the approach to risk by event organisers in a developed industry and those who perform in an emerging construct – their levels of risk acceptance, and how organisers in the developed and developing industries learn from the experiences of tragedy and new and changing legislations so that they are compliant.

Events are one-off live happenings which involve the management of short-term risks from event to event. This means organisers of events are less inclined to adopt strategic risk management behaviours and practices, whereas most risk cultures view risk for the longevity and long-term sustainability of an organisation or society.

Events are a service rather than a product because services must be experienced to be consumed. It means delivery and consumption are inseparable and makes it difficult to recover anything which goes wrong. Also, there is growing consumer demand for events to fulfil the need for transcendence and escapism for intense emotionalism and unrestrained exhibitionism (Rojek, 2013). Pressure to meet these demands means more can go wrong.

Silvers (2008) acknowledges that ethical practices are based on precedent, custom, tradition, standards, and fairness. But what is standard and customary will vary in different locations with different cultures. Organisers in both the UK and Poland do not routinely set objectives for their events and none of the interviewed organisers in either the UK or Poland mentioned safety as an objective. Mostly, the focus of organisers for a successful event is on financial return rather than a risk-free and safe event. This is because events are organised for financial gain, first, and not for the safety priority, even though every event is for people. This finding may be important and valuable in a wider context, but it is not a gap between the developed industry and the emerging one: it occurs with both.

Neither is formal education a factor in risk awareness because all interviewees, except one, mentioned risk or safety during the interviews whether they were formally educated or not. The one organiser who failed to mention risk or safety does have formal education.

Where there *is* a gap between the developed and emerging industries is in the level of acceptance of the presence of risk at events. Organisers in the UK are more likely to understand risk and the risk management processes to mitigate risk from causing harm because they are more attuned to living in a risk society with an audit culture approach. It is the law, then, which drives organisers or the company they work for to identify the safety priority.

Currently, there is no cohesive recognition by the industry, or consumers, or the insurance industry to lead organisers towards a culture of learning. This means there is a fragmented approach in the behaviours and practices of organisers, such as setting KPIs and identifying the need for the production and reproduction of knowledge.

UK organisers benefit from operating in a developed community of practice (Brown and Stokes, 2021), which has shaped to support organisers with compliance; otherwise, they face the real risk of repercussions and litigation. Whereas, in Poland, organisers are largely autonomous and immune to formal management regimes (Thiel, 2012) because the threat of repercussions due to noncompliance is distant. It is less likely in Poland that an authoritative body will visit a live event, and the event will be gone the next day, anyway. This is shaping a culture that encourages organisers to be complacent or perform deviant behaviour where risk is either ignored or partially addressed but is not fully mitigated from the potential to cause harm or injury. Thus, the acceptance of risk at events is lower in the developed industry, whereas there is a higher level of accepting risk by organisers in the developing industry.

The notion of fate and leaving risk to chance, or God, in the devout Catholic society of Poland is a factor because the issue with risk acceptance is its normalisation. Vaughan (1996) revealed a culture of incremental descent into poor judgement and a pattern of normalisation in engineering risk assessments by NASA before the launch of Challenger, which resulted in disaster. There is danger of the same occurring in Poland if societal acceptance of risk becomes embedded in event culture due to, among other things, fate, chance, and relying on God for protection. If this should happen, it will create a localised tier of higher risk acceptance compared to known and accepted levels of lower risk acceptance in the global marketplace for international events. This is troubling because events are becoming riskier with clients demanding ever more creativity, consumers expecting greater and greater experiences, and the societal shift towards adrenalin events, challenge events, and deviant events.

This is at odds with how the interviewed organisers consider their top three roles as an organiser of events: most responded with subjective aesthetical or emotional considerations such as good feelings and people being happy. Others considered the measurable commercial targets. Just three out of the 24 interviewees mentioned safety within their top three roles – all three of whom are organisers in the developed industry.

Some organisers are deflecting their social responsibility of delivering a risk-free event to other collaborators in their art world such as the venue, the client, or a supplier. By absolving themselves of this responsibility, organisers will not learn from their mistakes, which Pearn, Mulrooney and Payne (1998) say is important in turning blame culture into gain culture. The impact of this is that foreign clients and international consumers will not be aware of their exposure to a higher level of risk acceptance and lower level of risk mitigation from one location to another: that they are closer to harm without knowing it (involuntary acceptance of risk). Organisers should be communicating risks to consumers to convert involuntary risk acceptance to voluntary risk acceptance (Douglas and Wildavsky, 1982).

There are other risks which may cause harm to the success of an event such as unfulfilled revenues, financial loss, poor service, dissatisfied consumers, poor levels of attendance, psychological loss, and reputational harm.

To close this gap, it will require closer international collaboration across the global marketplace for international events so that organisers in emerging events industries can learn from the mistakes already experienced by a developed industry and begin to mirror the shape of their approach to risk, acceptance of risk, and mitigation of risk.

Mostly, events are the same wherever they take place in the world, and the risks are the same, which means they are preventable if organisers adopt a self-learning environment to learn from mistakes which have already happened at events elsewhere. Thus, an international socio-professional construct would allow for the role of tragedy to shape an emerging events industry by learning from the disaster experiences of an industry which is further developed and has shaped itself around tragic experiences that have already happened. This would prevent organisers in an emerging industry such as Poland from experiencing painful tragedies by themselves, and would, of course, make events safer around the globe for event tourists, foreign clients, international consumers, and workers and would align organisers in a united approach to risk and safety and reduce that gap.

If there were international collaboration and cross-cultural policy transfer learning, to raise the need for the production and reproduction of knowledge, an emerging events industry could shape itself around the premise of developing a culture of self-learning. Here, they would learn from other organisers by adopting legislations and practices that have come about because of tragedies already experienced and have shaped a developed industry. Such examples include the Bradford fire in 1985; the Hillsborough stadium crush in 1989; and the Manchester Arena bombing in 2017. Each of these tragedies at events resulted in inquiries that led to safer practices driven by legislations from the lessons learnt.

5 Author's Notes on Conclusions

'Event tourism' in event literature is the accepted term for consumers who travel to attend events. It is a generic term which blurs with *organisers* who travel to place their events in a foreign location, destination, or venue. To categorise organisers as 'client tourists' and an individual stakeholder group would be clearer and will identify their specific needs and expectations in terms of service, quality, safety, and for targeted marketing and promotion initiatives. If this distinction were to be adopted in event literature, it would help in events education for learning about the needs and expectations of client tourists and how to target and penetrate that specific market.

*

Having lived and worked as an organiser of events in both London and Warsaw for ten years, respectively, I had identified cultural and social differences, or 'gaps', in the approach to organising events, which have impacts for event tourists and client tourists in areas of service delivery, quality, and safety.

My privileged status (white, male, with Bachelor of Arts degree, royal and celebrity client portfolio) provided me with access to the socio-professional construct in both London and Warsaw without delay, question, or difficulty. I acknowledge that my status and career profile is unique in the events industry and the outcomes presented in this book will be different from a researcher with other ethnicity, gender, or career pathway.

I found, unsurprisingly, omissions in event practices and procedures in both the developed and developing constructs. These are: the failure to routinely set event objectives for every event to measure the success of the event against reaching those objectives; failure to conduct after-event feedback and evaluation procedures to understand if the objectives were met and obtain learning from feedback; and failure to always conduct risk management

DOI: 10.4324/9781032684161-5

procedures to identify, assess, and mitigate risks. These failures are replicated in both the UK and Poland events industries and so are not a gap between a developed and an emerging events industry. It is an important finding, nonetheless, because it reveals that the cause of these omissions is that the shape of an industry does not facilitate, encourage, or demand a culture of self-learning where organisers learn from others.

The events industry can learn from this book to prevent the international events industry from continuing to provide disparate practices and variants depending on where an event is happening and how the events industry in any location has shaped or is shaping due to its stage of development. To this end, there is need for an international umbrella association or construct to facilitate a recognised professional accreditation to align events industries worldwide to a cohesive global narrative.

This would enable organisers working within less mature events industries to rapidly catch up with a further developed events industry; to understand how to reach and attract foreign clients and international consumers; to narrow the gap between consumer expectations and the actual experience they receive; to learn from mistakes and crises that have already happened elsewhere so that they do not get replicated in different locations; to adopt legislations to ensure that organisers are compliant with risk management and safety protocols; and to enact best practices and procedures such as setting event objectives and carrying out after-event evaluation.

The intention would be to standardise international events so that foreign clients and international consumers receive a more consistent level of event management and safety wherever they are in the world.

Currently, the events industry is counterfactual from this book's contribution to research and industry. The events industry in the UK is disparate to that in Poland. It must follow, therefore, that all events industries in all locations have 'gaps' because each is at a different stage of development and each has sociocultural idiosyncrasies which shape their industry in a unique (thus disparate) way.

Academic courses in events management must begin to recognise and adopt cross-cultural content, with the purpose to educate students in understanding the factors which shape an industry and how they can (and must) implement their own self-learning environment. This is an imperative piece of events management education.

Researchers and academic authors in the discipline of events can respond by beginning to diversify their own research methodologies from the current domination of single case studies and instrumentalist approaches, and move towards cross-cultural research. In my novel methodological approach, I have shown that there is much to be contributed from qualitative autoethnography and great potential for leading academic authors to place themselves into their research by drawing on their own unique experiences and interpretations, as I have done. This would expand literature beyond its current limitations and provide a whole new range of theoretical studies from other researchers' perspectives alongside mine.

Dr Philip Berners
London, 2024

Bibliography

Abbott, A. (1988) *The system of the professions*. London: University of Chicago Press.

ACAS. (2021) *Putting someone on furlough: furlough and the Coronavirus job retention scheme*. Available at: www.acas.org.uk/coronavirus/furlough-scheme-pay (Accessed: 21 July 2021).

Adams, J. (2001) *Risk*. 5th edn. Abingdon: Routledge.

AEO. (2017) Available at: www.aeo.org.uk/ (Accessed: 13 November 2020).

Allen, J. (2009) *Event planning: the ultimate guide to successful meetings, corporate events, fund-raising galas, conferences, conventions, incentives and other special events*. Canada: J. Wiley & Sons. Available at: https://rl.talis.com/3/essex/lists/344A81EC-6735-B282-47A8-B84C10F2A222.html (Accessed: 27 November 2018).

Allen, J. *et al.* (2010) *Festival and special event management*. 5th edn. Brisbane: Wiley Australia.

Allen, P.M., Alston, F.E. and DeKerchove, E.M. (2019) *Peak performance*. 1st edn. Boca Raton: CRC Press, Taylor & Francis. https://doi.org/10.1201/9780429451508.

Andersen, T.J., Garvey, M. and Oliviero, R. (2014) *Managing risk and opportunity*. 1st edn. Oxford: Oxford University Press.

Anfara, V. and Mertz, N. (2011) *Mary Douglas's typology of grid and group, theoretical frameworks in qualitative research*. London: SAGE Publications, Inc. https://doi.org/10.4135/9781412986335.n8.

Argote, L. *et al.* (2000) 'Knowledge transfer in organizations: learning from the experience of others', *Organizational Behavior and Human Decision Processes*, 82(1), pp. 1–8. https://doi.org/10.1006/obhd.2000.2883.

Armstrong, G., Giulianotti, R. and Hobbs, D. (2017) *Policing the 2012 London Olympics: legacy and social exclusion*. Abingdon: Routledge.

Armstrong, G., Hobbs, D. and Lindsay, I. (2011) 'Calling the shots: the pre-2012 London Olympic contest', *Urban Studies*, 48(15). Available at: https://journals.sagepub.com/doi/full/10.1177/0042098011422397.

Augustyn, M. and Ho, S.K. (1998) 'Service quality and tourism', *Journal of Travel Research*, 37(1). https://doi.org/10.1177/004728759803700110.

Back, L. (2007) *The art of listening*. Oxford: Berg.

Badawy, M. *et al.* (2016) 'A survey on exploring key performance indicators', *Future Computing and Informatics Journal*, 1(1–2), pp. 47–52. https://doi.org/10.1016/j.fcij.2016.04.001.

Baggs, M. (2019) 'Fyre festival: inside the world's biggest festival flop', *BBC News*. Available at: www.bbc.co.uk/news/newsbeat-46904445 (Accessed: 15 January 2022).

Banks, E. (2012) *Risk culture*. Basingstoke: Palgrave Macmillan.

Bauman, Z. (2000) *Liquid modernity*. Oxford: Blackwell Publishers Ltd.

BBC News. (2019) Available at: www.bbc.co.uk/news/world-europe-46765692 (Accessed: 4 February 2019).

Bebko, C.P. (2000) 'Service intangibility and its impact on consumer expectations of service quality', *Journal of Services Marketing*, 14(1). Available at: www.emerald.com/insight/content/doi/10.1108/08876040010309185/full/pdf?title=service-intangibility-and-its-impact-on-consumer-expectations-of-service-quality (Accessed: 26 April 2021).

Beck, U. (2007) *World at risk*. Cambridge: Polity Press.

Becker, H.S. (1928/2008) *Art worlds/Howard S. Becker*. Berkeley, CA: University of California Press. Available at: https://encore.essex.ac.uk/iii/encore/record/C__Rb1664558__Sbecker art__Orightresult__U__X2?lang=eng&suite=cobalt (Accessed: 27 November 2018).

Becker, H.S. (1984) *Art worlds*. 1st edn. London: University of California Press.

Becker, H.S. (2007) *Telling about society*. London: University of Chicago Press.

Beckford, J. (2017) *Quality – a critical introduction*. 4th edn. Abingdon: Routledge.

Beech, J.G., Kaiser, S. and Kaspar, R. (2014) *The business of events management*. Harlow: Pearson Education. https://doi.org/10.1073/pnas.1011284108.

Bennett, T., Grossberg, L. and Morris, M. (2005) *New keywords: a revised vocabulary of culture and society*. Oxford: Blackwell Publishing Ltd.

Berners, P. (2013) *Happy event*. Warsaw: Priorytet.

Berners, P. (2017) *The practical guide to organising events*. Abingdon: Routledge.

Berners, P. (2019) *The practical guide to managing event venues*. Abingdon: Routledge.

Berners, P. and Martin, A. (2022) *The practical guide to achieving customer satisfaction in events and hotels*. Abingdon: Routledge.

Berridge, G. (2010) 'Event pitching: the role of design and creativity', *International Journal of Hospitality Management*, 29, pp. 208–215.

Berridge, G. (2011) *Events design and experience*. 2nd edn. Abingdon: Routledge.

Berridge, G. (2012) 'Event experience: a case study of differences between the way in which organisers plan an event experience and the way in which guests receive the experience', *Journal of Park and Recreational Administration*, 30(3), pp. 7–23.

'Beyonce won't visit us anymore?'. (2013) Poland: Dzien Dobry TVN. Available at: https://dziendobry.tvn.pl/gwiazdy/beyonce-juz-nas-nie-odwiedzi-da283211.

Bladen, C. *et al.* (2018) *Events management: an introduction*. 2nd edn. Abingdon: Routledge.

Bogdan, W. *et al.* (2015) *Poland 2025: Europe's new growth engine*. Available at: www.mckinsey.pl (Accessed: 1 May 2021).

Boswijk, A., Thijssen, T. and Peelen, E. (2007) *The experience economy: a new perspective*. Amsterdam: Pearson Education Benelux.

Bourdieu, P. (1990) *The logic of practice*. Cambridge: Stanford University Press.

Bowdin, G. *et al.* (2011) *Events management*. 3rd edn. Abingdon: Routledge.

Boyne, R. (2003) *Risk*. Buckingham: Open University Press.

Brown, D.K. (2001) 'The social sources of educational credentialism: status cultures, labor markets, and organizations', *Sociology of Education*, 74, pp. 34–89.

Brown, T. and Stokes, P. (2021) 'Events management as a community of practice', *Journal of Hospitality and Tourism Insights*, 4(1). https://doi.org/10.1108/JHTI-09-2020-0157.

BSI. (2020) Available at: www.bsigroup.com/en-GB/ (Accessed: 15 November 2020).

Buczak, T. *et al*. (2020) *Poland meetings and events industry report*. Torun. Available at: www.pot.gov.pl (Accessed: 9 August 2021).

Buhalis, D. (2000) 'Marketing the competitive destination of the future', *Tourism Management*, 21(1), pp. 97–116. https://doi.org/10.1016/S0261-5177(99)00095-3.

Buras, P. (2019) 'The killing of Gdansk's mayor is the tragic result of hate speech', *The Guardian*, January. Available at: www.theguardian.com/commentisfree/2019/jan/17/gdansk-mayor-pawel-adamowicz-killing-poland.

Burke, R.J. (1995) 'Benefits of formal training courses within a professional services firm', *Journal of Management Development*, 14(3), pp. 3–13. https://doi.org/10.1108/02621719510073920

Business Visits and Events Partnership. (2020) *BVEP*. Available at: www.businessvisitsandeventspartnership.com/research-and-publications/research-directory/1534-the-uk-events-report-full-report-1 (Accessed: 1 August 2021).

Cairncross, F. (2001) *The death of distance*. London: Texere Publishing Limited.

Cambridge University Press. (2021) *LANDMARK | meaning in the Cambridge English Dictionary, Cambridge Dictionary*. Available at: https://dictionary.cambridge.org/dictionary/english/landmark (Accessed: 16 July 2021).

Cambridge English Dictionary (2024) Available at: https://dictionary.cambridge.org (Accessed 9 March 2024).

Carrabine, E. (2007) 'Michel Foucault', in Scott, J. (ed.) *Fifty key sociologists*. Abingdon: Routledge, pp. 81–87.

Carson, D. and Bain, A. (2008) *Professional risk and working with people*. London: Jessica Kingsley Publishers.

CDM. (2015) Available at: www.hse.gov.uk/construction/cdm/2015/index.htm (Accessed: 16 November 2020).

Chang, H. (2008) *Autoethnography as method*. Abingdon: Routledge.

Chartered Institute of Marketing | CIM. (2021) Available at: www.cim.co.uk/?gclid=Cj0KCQjwppSEBhCGARIsANIs4p5Q9U-5GwV__e-kpiUFc9jStzv98CRdVaevvakP-aGFGY9vfrchg80aAgW6EALw_wcB&gclsrc=aw.ds (Accessed: 25 April 2021).

Cinecitta. (2021) Available at: https://cinecittastudios.it/ (Accessed: 4 April 2022).

CIPD. (2021) *Coronavirus (COVID-19): furlough guide*. Available at: www.cipd.co.uk/knowledge/fundamentals/emp-law/employees/furlough#gref (Accessed: 21 July 2021).

Clasen, M., Andersen, M. and Schjoedt, U. (2019) 'Adrenaline junkies and white-knucklers: a quantitative study of fear management in haunted house visitors', *Poetics*, 73, pp. 61–71. https://doi.org/10.1016/j.poetic.2019.01.002.

Clegg, S.R., Skyttermoen, T. and Vaagaasar, A.L. (2021) *Project management: a value creation approach*. London: SAGE Publications Ltd.

The Concise Oxford Dictionary of Sociology. (1994) Oxford: Oxford University Press.

Conference & Incentive Travel. (2017) *State of the industry report*. Available at: www.citmagazine.com/industry-reports (Accessed: 24 May 2019).

Conference & Incentive Travel. (2018) *State of the industry report, conference & incentive travel*. Available at: www.citmagazine.com/industry-reports (Accessed: 24 May 2019).

Conference News. (2019) 'London leading the pack', January, pp. 8–9.

Cooper, C. (2008) *Tourism principles and practice.* 4th edn. Harlow: Pearson Education.

Corporate Events Packages & Facilities | Thorpe Park Resort. (2021) Available at: www.thorpepark.com/corporate-events/ (Accessed: 29 June 2021).

Cottle, S. (no date) 'Ulrich Beck, 'risk society' and the media', *European Journal of Communication.* Available at: https://journals.sagepub.com/doi/pdf/10.1177/026732 3198013001001 (Accessed: 9 December 2020).

Crompton, J. and McKay, S. (1997) 'Motives of visitors attending festival events', *Annals of Tourism Research,* XXIV(2), pp. 425–439.

Crook, T. and Esbester, M. (2016) *Governing risks in modern Britain.* London: Macmillan Publishers Ltd.

The Crystal Maze. (2020) Available at: https://the-crystal-maze.com/ (Accessed: 13 November 2022).

Cserháti, G. and Szabó, L. (2014) 'The relationship between success criteria and success factors in organisational event projects', *International Journal of Project Management,* 32(4), pp. 613–624. https://doi.org/10.1016/j.ijproman.2013.08.008.

Curran, J. and Morley, D. (2006) *Media and cultural theory.* Abingdon: Routledge.

Data Protection Act 2018. (2021) Available at: www.legislation.gov.uk/ukpga/2018/12/contents/enacted (Accessed: 22 July 2021).

Davies, W. (2017) 'Rio Olympics 2016: should Brazil have done better with Olympic legacy?', *BBC Sport.* Available at: www.bbc.co.uk/sport/olympics/39323546 (Accessed: 21 May 2019).

de Kadt, E. and Williams, G. (eds.) (1976) *Sociology and development.* 2nd edn. London: Tavistock Publications Limited.

Deery, M. and Jago, L. (2010) *Social impacts of events and the role of anti-social behaviour.* Available at: www.emeraldinsight.com/doi/full/10.1108/17852951011029289 (Accessed: 14 October 2018).

Desjardins, A. *et al.* (2021) 'First-person methods in HCI', *ACM Transaction on Computer-Human Interaction,* 28(6), pp. 1–12. Article No. 37. Available at: https://0-dl-acm-org.serlib0.essex.ac.uk/doi/pdf/10.1145/3492342.

Desmond, J. (1999) *Staging tourism: bodies on display from Waikiki to sea world.* London: The University of Chicago Press.

Dey, I. (1993) *Qualitative data analysis.* Abingdon: Routledge.

Dobosz-Bourne, D. (2004) *Knowledge transfer across cultural boundaries in the global economy based on the model of travel of ideas exemplified by the quality transfer in car manufacture from West Europe to Poland.* Luton: University of Bedfordshire.

Dolowitz, D. and Marsh, D. (1996) 'Who learns what from whom: a review of the policy transfer literature', *Political Studies Association,* 44, pp. 343–357.

Domański, B. (2003) 'Industrial change and foreign direct investment in the postsocialist economy the case of Poland', *European Urban and Regional Studies,* 10(2), pp. 99–118. Available at: www.sagepublications.com (Accessed: 20 August 2021).

Domanski, H. (1999) 'East central Europe', *East Central Europe International Journal of the Social Sciences and Humanities,* 26(1).

Donthu, N. and Yoo, B. (1998) 'Cultural influences on service quality expectations', *Journal of Service Research,* 1(2), pp. 178–186. https://doi.org/10.1177/109467059800100207.

Douglas, M. (2003) *Risk and blame.* Abingdon: Routledge.

Douglas, M. and Wildavsky, A. (1982) *Risk and culture: an essay on the selection of technological and environmental dangers.* London: University of California Press.

Durkheim, E. (1952) *Suicide: a study in the sociology.* Glencoe, IL: Free Press.

Edge Hotel School | University of Essex. (2021) Available at: www.essex.ac.uk/depart ments/edge-hotel-school (Accessed: 19 April 2021).

Ekberg, M. (2007) 'The parameters of the risk society', *Current Sociology,* 55(3), pp. 343–366. https://doi.org/10.1177/0011392107076080.

EMBOK. (no date) Available at: www.embok.org/index.php/embok-model (Accessed: 1 February 2019).

Evans, M. (2004) *Policy transfer in global perspective.* Abingdon: Routledge.

Evans, M. and Davies, J. (1999) 'Understanding policy transfer: a multi-level, multi-disciplinary perspective', *Public Administration,* 77(2), pp. 361–385. https://doi. org/10.1111/1467-9299.00158.

Evans, N. (2015) *Strategic management for tourism, hospitality and events.* 2nd edn. Abingdon: Routledge.

Eyal, G. and Pok, G. (no date) *From a sociology of professions to a sociology of expertise.* Columbia University, New York, NY, USA.

Ferdinand, N. and Kitchin, P. (2017) *Events management: an international approach.* 2nd edn. London: SAGE Publications Ltd.

FIFA. (2014) '2014 FIFA World Cup Brazil™ – matches', *FIFA.com.* Available at: www.fifa.com/worldcup/archive/brazil2014/matches/ (Accessed: 21 May 2019).

Frankel, J. (1979) *International relations in a changing world.* Oxford: Oxford University Press.

Gallagher, B. (2018) 'Nicki Minaj cancels Shanghai concert that was a "scam" ', *Daily Mail.* Available at: www.dailymail.co.uk/tvshowbiz/article-6416033/Nicki-Minaj-cancels-Shanghai-concert-scam.html (Accessed: 21 August 2021).

Galloway, S. (2006) 'Adventure recreation reconceived: positive forms of deviant leisure', *Leisure/Loisir,* 30(1), pp. 219–231. https://doi.org/10.1080/14927713.2006.9651349.

Gapinski, J. (2001) 'The production of culture', *Review of Economics and Statistics,* 62(4), pp. 578–586.

Gardham, D. (2020) 'Manchester bombing: police unaware Ariana Grande concert was taking place on night of attack', *Sky News,* 8 September. Available at: https://news. sky.com/story/manchester-bombing-police-unaware-ariana-grande-concert-was-taking-place-on-night-of-attack-inquiry-hears-12066576.

Gelder, G. and Robinson, P. (2009) 'A critical comparative study of visitor motivations for attending music festivals: a case study of glastonbury and V festival', *Event Management,* XIII(3), pp. 181–196.

Gellner, E. (1988) *Plough, sword and book.* London: Collins Harvill.

Gerber, M. and Von Solms, R. (2005) 'Management of risk in the information age', *Computers & Security,* 24(1), pp. 16–30. https://doi.org/10.1016/J.COSE.2004.11.002.

Gerritsen, D. and Van Olderen, R. (2014) *Events as a strategic marketing tool.* Wallingford: CAB International.

Getz, D. (2018) *Event evaluation.* Oxford: Goodfellow Publishers Ltd.

Getz, D. and Page, S.J. (2016) *Event studies.* 3rd edn. Abingdon: Routledge.

Giddens, A. (2004) *Sociology.* 4th edn. Cambridge: Blackwell Publishers Ltd.

Giddens, A. (2012) *The consequences of modernity.* 19th edn. Cambridge: Polity Press.

Gillespie, K. and Hennessey, D. (2016) *Global marketing.* 4th edn. Abingdon: Routledge.

Giulianotti, R. (2004) *Sport and modern social theorists.* Basingstoke: Palgrave Macmillan.

Giulianotti, R. *et al.* (2015) 'Sport mega-events and public opposition: a sociological study of the London 2012 Olympics', *Journal of Sport and Social Issues*, 39(2), p. 99.

Globex. (2018) 'Emerging markets to grow faster than mature regions', *Exhibition World*, p. 17.

Goble, R., Bier, V. and Renn, O. (2018) 'Two types of vigilance are essential to effective hazard management: maintaining both together is difficult', *Risk Analysis*, 38(9), pp. 1795–1801. https://doi.org/10.1111/risa.13003.

Goldblatt, J.J. (2011) *Special events*. 6th edn. New York: Wiley.

Goldblatt, J.J. (2014) *Special events: creating and sustaining a new world for celebration*. 7th edn. Hoboken, NJ: Wiley.

Gomułka, S. (2016) 'Poland's economic and social transformation 1989–2014 and contemporary challenges', *Central Bank Review*, 16(1), pp. 19–23. https://doi.org/10.1016/j.cbrev.2016.03.005.

gov.uk. (2020) *Coronavirus (COVID-19)*. Available at: www.gov.uk/coronavirus (Accessed: 29 September 2020).

Gov.uk (2022) Available at: https://www.gov.uk/government/news/martyns-law (Accessed: 14 March, 2024).

Greene, M. (1983) *Marketing hotels into the nineties*. 1st edn. William Heinemann.

Grenfell, M. (ed.) (2012) *Pierre Bourdieu key concepts*. 2nd edn. Durham: Acumen Publishing Ltd.

Gronroos, C. (1988) 'Service quality: the six criteria of good perceived service – ProQuest', *Review of Business*, 9(3). Available at: https://search.proquest.com/openview/a4947917a28900d240398317bd492ac9/1?pq-origsite=gscholar&cbl=36534 (Accessed: 27 April 2021).

Gruneau, R.S. and Horne, J. (2016) 'Mega-events and globalization: capital and spectacle in a changing world order', pp. 44–46. Available at: https://books.google.co.uk/books?id=bv2oCgAAQBAJ&pg=PA45&lpg=PA45&dq=rpts+tamu+faculty+Economic+Impact&source=bl&ots=MO-Zs06lxF&sig=ESEFZTFJ9w_cMYETFmBJtz4fwNU&hl=en&sa=X&ved=2ahUKEwji2ba1vZXfAhVVSBUIHbCqA7wQ6AEwBHoECAMQAQ#v=onepage&q=rpts tamu faculty Ec (Accessed: 10 December 2018).

The Guardian. (no date) 'Revealed: 6,500 migrant workers have died since World Cup awarded'.

Gutting, G. (ed.) (2005) *The Cambridge companion to Foucault*. 2nd edn. Cambridge: Cambridge University Press.

Halliday, T.C. (1987) *Beyond monopoly*. London: University of Chicago.

Hammersley, M. and Atkinson, P. (2007) *Ethnography: principles in practice*. London: Tavistock Publications Limited.

Haniff, A. and Salama, M. (2016) *Project management*. Oxford: Goodfellow Publishers Ltd.

Hanquinet, L. and Savage, M. (eds.) (2016) *Routledge international handbook of the sociology of art and culture*. Abingdon: Routledge.

Harker, R., Mahar, C. and Wilkes, C. (eds.) (1990) *An introduction to the work of Pierre Bourdieu*. New York: St Martin's Press.

Hayano, D.M. (1979) 'Auto-ethnography: paradigms, problems and prospects', *Society for Applied Anthropology*, 38(1), pp. 99–104.

Hays, S., Page, S.J. and Buhalis, D. (2013) 'Social media as a destination marketing tool: its use by national tourism organisations', *Current Issues in Tourism*, 16(3), pp. 211–239. https://doi.org/10.1080/13683500.2012.662215.

Hayward, K. and Smith, O. (2017) *The Oxford handbook of criminology.* 6th edn. Edited by A. Liebling, S. Maruna and L. McAra. Oxford: Oxford University Press.

Hede, A.-M. (2008) 'Managing special events in the new era of the triple bottom line', *Event Management*, 11(1–2), pp. 13–22.

Hillson, D. and Murray-Webster, R. (2007) *Understanding and managing risk attitude.* 2nd edn. Aldershot: Gower Publishing Ltd.

Holmes, K. and Ali-Knight, J. (2017) 'The event and festival life cycle – developing a new model for a new context', *International Journal of Contemporary Hospitality Management*, 29(3), pp. 986–1004. https://doi.org/10.1108/IJCHM-10-2015-0581.

Home – Hofstede Insights. (no date) Available at: www.hofstede-insights.com/ (Accessed: 11 April 2019).

Horne, J. and Manzenreiter, W. (2006) 'An introduction to the sociology of sports mega-events ', *The Sociological Review*, 54(2_suppl), pp. 1–24. https://doi.org/10.1111/j.1467-954X.2006.00650.x.

Hospers, G.-J. (2004) 'Place marketing in Europe', *Intereconomics*, 39(5), pp. 271–279. https://doi.org/10.1007/bf03031785.

Howard, D.A. (2006) *Poland's transformation a work in progress.* Edited by M.J. Chodakiewicz, J. Radzilowski and D. Tolczyk. London: Transaction Publishers.

HSE. (no date) *Managing risk assessments.* Available at: www.hse.gov.uk/simple-health-safety/risk/risk-assessment-template-and-examples.htm (Accessed: 18 November 2020).

ICA. (no date) Available at: www.int-comp.org/ (Accessed: 13 November 2020).

International Confex – Where the events industry meets. (2021) Available at: www.international-confex.com/welcome?utm_source=google&utm_medium=cpc&utm_campaign=tagdigital_confex_brand&gclid=Cj0KCQjwvO2IBhCzARIsALw3A-SrWBaWrJ86bTgGCuXcG5cukZlLFs73WMckTjkhlM6XbwD1KSDDQ1DwaAhOGEALw_wcB (Accessed: 17 August 2021).

Jackson, C., Morgan, J. and Laws, C. (2018) 'Creativity in events: the untold story', *International Journal of Event and Festival Management*, 9(1).

Jago, L.K. and Shaw, R.N. (1998) 'Special events: a conceptual and definitional framework', *Festival Management and Event Tourism*, 5(1), pp. 21–32. https://doi.org/10.3727/106527098792186775.

Jeston, J. (2018) *Business process management.* Abingdon: Routledge.

Kahle, L.R. and Close, A.G. (eds.) (2011) *Customer behaviour knowledge for effective sports and event marketing.* Hove: Taylor and Francis Group.

Kavaratzis, M. (2004) 'From city marketing to city branding: towards a theoretical framework for developing city brands', *Place Branding*, 1(1), pp. 58–73. https://doi.org/10.1057/palgrave.pb.5990005.

Kelly, G. (1991) *The psychology of personal constructs.* Routledge in Association with the Centre for Personal Construct Psychology. Available at: https://books.google.co.uk/books/about/The_Psychology_of_Personal_Constructs.html?id=z8tJ92sBE9cC&printsec=frontcover&source=kp_read_button&redir_esc=y#v=onepage&q&f=false (Accessed: 12 December 2018).

Kemp-Welch, A. (2008) *Poland under communism a cold war history.* Cambridge: Cambridge University Press.

Kielbasiewicz-Drozdowska, I. (2005) 'Leisure time as space for gaining social capital', *Studies in Physical Culture and Tourism*, 12(1). Available at: www.wbc.poznan.pl/Content/21056/XSL Output.html#IDACBU1B (Accessed: 17 May 2021).

Kim, K., Uysal, M. and Chen, J.S. (2014) 'Festival visitor motivation from the organizers' points of view', *Event Management*, 7(2), pp. 127–134. https://doi.org/10.3727/152599501108751533.

Kim, N.-S. and Chalip, L. (2004) 'Why travel to the FIFA World Cup? Effects of motives, background, interest, and constraints', *Tourism Management*, 25(6), pp. 695–707. https://doi.org/10.1016/j.tourman.2003.08.011.

King, N. and Horrocks, C. (2012) *Interviews in qualitative research*. 2nd edn. London: SAGE Publications Ltd.

Klegon, D. (1978) The Sociology of Professions: An Emerging Perspective. *sociology of work and occupations*, 5(3), pp. 259–283. https://doi.org/10.1177/07308884780500301

Koffman, J. *et al.* (2020) 'Uncertainty and COVID-19: how are we to respond?', *Journal of the Royal Society of Medicine*, 113(6), pp. 211–216. https://doi.org/10.1177/0141076820930665.

Kuhlmann, E. (2013) 'Sociology of professions: towards international context-sensitive approaches', *South African Review of Sociology*, 44(2), pp. 7–17. https://doi.org/10.1080/21528586.2013.802534.

Lakshmi, S. and Mohideen, M.A. (2012) 'Issues in reliability and validity of research', *International Journal of Management Research and Review*, 3(4). Available at: www.ijmrr.com (Accessed: 21 July 2021).

Lane, T. and Wolanski, M. (2009) *Poland and European integration*. Basingstoke: Palgrave Macmillan.

Larson, M.S. (1977) *The rise of professionalism: a sociological analysis*. London: University of California Press.

Leavis, F.R. (1952) *The common pursuit*. London: Chatto and Windus.

Lena, J.C. (2015) 'The production of culture: prospects for the twenty-first century', *Sociology*, pp. 608–613.

Lenski, G., Nolan, P. and Lenski, J. (2014) 'Human societies: an introduction into macro sociology', *Open Journal of Social Sciences*, 2(8). Available at: www.scirp.org/(S(i43dyn45teexjx455qlt3d2q))/reference/ReferencesPapers.aspx?ReferenceID=1272375 (Accessed: 24 April 2021).

Levy, C., Lamarre, E. and Twining, J. (2010) *Taking control of organizational risk culture risk practice McKinsey working papers on risk*. Available at: www.mckinsey.com/~/media/mckinsey/dotcom/client_service/risk/working papers/16_taking_control_of_organizational_risk_culture.ashx (Accessed: 5 March 2019).

Lipton, D. *et al.* (1990) 'Creating a market economy in eastern Europe: the case of Poland', *Source: Brookings Papers on Economic Activity*, 1990(1), pp. 75–147. Available at: www.jstor.org/stable/2534526 (Accessed: 11 February 2021).

Lynn, K. (1963) 'Introduction to the professions', *Daedalus*, Fall.

Macdonald, K.M. (1999) *The sociology of the professions*. 2nd edn. London: SAGE Publications Ltd.

Macionis, J. and Plummer, K. (1998) *Sociology a global introduction*. 6th edn. Hoboken, NJ, USA. Prentice Hall.

Mackellar, J. (2014) *Event audiences and expectations*. Abingdon: Routledge.

Mannin, M. (ed.) (1999) *Pushing back the boundaries the European Union and Central and Eastern Europe*. Manchester: Manchester University Press.

Mash Media. (2019) 'A global player', *Exhibition News*, January, p. 27.

Mason, P. (2014) *Researching tourism, leisure and hospitality for your dissertation.* Oxford: Goodfellow Publishers Ltd.

McCabe, S., Sharples, M. and Foster, C. (2012) 'Stakeholder engagement in the design of scenarios of technology-enhanced tourism services', *Tourism Management Perspectives*, 4, pp. 36–44. https://doi.org/10.1016/j.tmp.2012.04.007.

Mehta, J. *et al.* (2020) 'Rapid implementation of Microsoft teams in response to COVID-19: one acute healthcare organisation's experience', *BMJ Health and Care Informatics*, 27(3). https://doi.org/10.1136/bmjhci-2020-100209.

Merriam-Webster. (2021) *Definition of landmark by Merriam-Webster, Dictionary, Merriam-Webster.* Available at: www.merriam-webster.com/dictionary/landmark (Accessed: 16 July 2021).

MICE Meetings Incentives Conferences Events Incoming Tour Operator. (2018) Available at: www.visitpoland.com/mice (Accessed: 14 March 2021).

Mills, C.W. (1956) *White collar.* New York: Oxford University Press.

Mills, C.W. (1959) *The power elite.* New York: Oxford University Press.

Mishler, W. and Rose, R. (1997) 'The journal of politics', *Journal of Politics*, 59(2), pp. 418–451. Available at: https://arizona.pure.elsevier.com/en/publications/trust-distrust-and-skepticism-popular-evaluations-of-civil-and-po (Accessed: 21 May 2019).

Moss-Kanter, R. (1983) *The change masters.* New York: Simon and Schuster.

Moufakkir, O. and Pernecky, T. (eds.) (2015) *Ideological, social and cultural aspects of events.* Wallingford: CAB International.

Mules, T. (2004) 'Case study evolution in event management: the gold coast's wintersun festival', *Event Management*, 9(1), pp. 95–101. https://doi.org/10.3727/1525995042781075.

Muncey, T. (2010) *Creating autoethnographies.* London: SAGE Publications Ltd.

Murchison, J.M. (2010) *Ethnography essentials.* San Francisco: John Wiley and Sons Inc.

NEBOSH. (no date) Available at: www.nebosh.org.uk/home/ (Accessed: 13 November 2020).

The O2. (2021) Available at: www.theo2.co.uk/ (Accessed: 15 August 2021).

O'Donnell, M. (1997) *Introduction to sociology.* 4th edn. Walton-on-Thames: Thomas Nelson and Sons Ltd.

Olympia London. (2018) Available at: https://olympia.london/ (Accessed: 15 August 2021).

Orange Warsaw Festival. (2013) Available at: https://orangewarsawfestival.pl/en/o-festiwalu/2013 (Accessed: 14 March 2021).

Outhwaite, W. and Ray, L.J. (2005) *Social theory and postcommunism.* Oxford: Blackwell Publishers Ltd.

Pałac Kultury i Nauki. (2021) Available at: https://pkin.pl/en/home/ (Accessed: 15 August 2021).

Paraskevas, A. (2006) 'Crisis management or crisis response system?', *Management Decision*, 44(7), pp. 892–907. Edited by R. Wilding. https://doi.org/10.1108/00251740610680587.

Parasuraman, A., Zeithaml, V.A. and Berry, L.L. (1988) 'SERVQUAL: a multiple-item scale for measuring consumer perceptions of service quality', *Journal of Retailing*, 64(1).

Pearn, M., Mulrooney, C. and Payne, T. (1998) *Ending the blame culture.* Aldershot: Gower Publishing Ltd.

Peterson, R.A. and Anand, N. (2004) 'The production of culture perspective', *Annual Review of Sociology*, 30(1), pp. 311–334. https://doi.org/10.1146/annurev. soc.30.012703.110557.

PGE Narodowy – Warsaw National Stadium – The Stadium Guide. (no date) Available at: www.stadiumguide.com/stadionnarodowy/ (Accessed: 17 March 2021).

Pielichaty, H. *et al*. (2017) *Events project management*. 1st edn. Abingdon: Routledge.

Pine, B.J. and Gilmore, J.H. (2011) *The experience economy*. 2nd edn. Boston MA: Harvard Business Review Press.

Plummer, K. (2019) *Narrative power: the struggle for human value*. Cambridge: Polity Press.

Poland Convention Bureau. (2020) Available at: www.pot.gov.pl/en/poland-convention (Accessed: 15 November 2020).

Poland | European Commission. (no date) Available at: https://ec.europa.eu/info/busi ness-economy-euro/economic-performance-and-forecasts/economic-performance-country/poland_en (Accessed: 1 May 2021).

Poland Events Impact 2019 Report. (2021) Available at: www.pot.gov.pl/en/poland-convention/news/launch-of-the-poland-events-impact-2019-report (Accessed: 7 March 2021).

Poulsson, S.H.G. and Kale, S.H. (2004) 'The experience economy and commercial experiences', *The Marketing Review*, 4(3), pp. 267–277. https://doi.org/10.1362/ 1469347042223445.

Power, D. and Scott, A. (2004) *Cultural industries and the production of culture*. Abingdon: Routledge.

PRINCE2. (2020) Available at: www.prince2.com/uk/training/prince2 (Accessed: 15 November 2020).

Professional Convention Management Association (PCMA). (2020) Available at: www. hospitalitynet.org/organization/17007459/pcma.html (Accessed: 20 April 2021).

Pugh, C. and Wood, E.H. (2004) 'The strategic use of events within local government: a study of London borough councils', *Event Management*, 9, pp. 61–71.

The Purple Guide. (no date) Available at: www.thepurpleguide.co.uk/ (Accessed: 3 December 2019).

Qu, S.Q. and Dumay, J. (2011) 'The qualitative research interview', *Qualitative Research in Accounting & Management*, 8(3), pp. 238–264. Available at: www. emerald.com/insight/content/doi/10.1108/11766091111162070/full/pdf?title=the-qualitative-research-interview.

Quick, L. (2020) *Managing events*. London: SAGE Publications Ltd.

Quinn, B. (2013) *Key concepts in event management*. 1st edn. London: SAGE Publications Ltd.

Raj, R., Walters, P. and Rashid, T. (2015) *Events management principles and practice*. 3rd edn. London: SAGE Publications Ltd.

Regester, M. and Larkin, J. (2002) *Risk issues and crisis management*. London: Kogan Page Ltd.

Reic, I. (2017) *Events marketing management: a consumer perspective*. Abingdon: Routledge. Available at: http://encore.essex.ac.uk/iii/encore/record/C__Rb2060925__ Sreic__Orightresult__U__X4?lang=eng&suite=cobalt (Accessed: 25 November 2018).

Ritzer, G. and Walczak, D. (1986) *Working, conflict and change*. 3rd edn. Harlow: Longman Higher Education.

Robertson, R. (ed.) (2014) *European glocalisation in global context.* Basingstoke: Palgrave Macmillan.

Roche, M. (2000) *Megaevents and modernity.* Abingdon: Routledge.

Rojek, C. (2000) *Leisure and culture.* Basingstoke: Macmillan Press Ltd.

Rojek, C. (2013) *Event power: how global events manage and manipulate.* London: SAGE Publications Ltd.

Roncak, M. (2019) 'CMW', *Conference & Meetings World,* p. 61. Available at: www.c-mw.net/ (Accessed: 21 August 2021).

Rose, R. (1993) *Lesson-drawing in public policy: Aguide to learning across time and space (Vol. 91).* Chatham: Chatham House Publishers.

Rowe, D. (2017) 'Birmingham centre for contemporary cultural studies', in *The Wiley-Blackwell encyclopedia of social theory,* pp. 1–5. https://doi.org/10.1002/9781118430873.EST0575.

Ruane, J.M. (2016) *Social research methods.* Chichester: John Wiley and Sons Ltd.

Rubin, H.J. and Rubin, I.S. (2012) *Qualitative interviewing.* 3rd edn. London: SAGE Publications Ltd.

Salameh, R. (2019) 'Brazil reaching for its Olympic legacy', *Conference & Meetings World,* pp. 20–22.

Saunders, J. (2021) *Manchester Arena inquiry: volume 1 security for the Arena.* London: Her Majesty's Stationery Office.

Saunders, M., Lewis, P. and Thornhill, A. (2016) *Research methods for business students.* Harlow: Pearson Education Ltd.

Schulz, B. (2008) 'The importance of soft skills: education beyond academic knowledge', *Journal of Language and Communication.* pp. 146–155. Polytechnic of Namibia.

Schwarz, E.C. *et al.* (2017) *Managing sport facilities and major events.* 2nd edn. Abingdon: Routledge.

Scott, J. (1997) *Sociological theory contemporary debates.* Cheltenham: Edward Elgar Publishing Ltd.

Seidman, I. (2013) *Interviewing as qualitative research.* New York: Teachers College Press.

Shaw, C., Anin, R. and Vdovii, L. (2015) *Ghosts of Sochi: hundreds killed in Olympic construction, radio free Europe.* Available at: www.rferl.org/a/ghosts-of-sochi-olympics-migrant-deaths/26779493.html (Accessed: 27 January 2022).

Shils, E. (1961) 'Mass society and its culture', in Jacobs, N. (ed.) *Culture for the millions? Mass media in modern society.* Princeton, NJ: D Van Nostrand, pp. 1–27.

Shone, A. and Parry, B. (2013) *Successful event management: a practical handbook.* Cengage Learning. Available at: https://rl.talis.com/3/essex/lists/344A81EC-6735-B282-47A8-B84C10F2A222.html (Accessed: 27 November 2018).

Silvers, J.R. (2008) *Risk management for meetings and events.* Abingdon: Routledge.

Smaho, M. (2012) *System of knowledge transfer in the automotive industry.* Universitas-Gyor. Available at: file:///C:/Users/staff/Downloads/Smaho_2012_knowledgetransfer_automotive.pdf.

Smith, M.K. and Richards, G. (2013) *The Routledge handbook of cultural tourism.* 1st edn. Abingdon: Routledge.

Smith, M.K. and Robinson, M. (eds.) (2009) *Cultural tourism in a changing world.* 2nd edn. Clevedon: Channel View Publications.

Smith, P. and Riley, A. (2009) *Cultural theory: an introduction.* 2nd edn. Oxford: Blackwell Publishers Ltd.

Sönmez, S.F. and Graefe, A.R. (1998) 'Influence of terrorism risk on foreign tourism decisions', *Annals of Tourism Research*, 25(1), pp. 112–144. https://doi.org/10.1016/S0160-7383(97)00072-8.

Spillman, L. (ed.) (2002) *Cultural sociology.* Oxford: Blackwell Publishers Ltd.

Storey, J. (2009) *Cultural theory and popular culture.* 4th edn. Harlow: Pearson Education.

Strathern, M. (ed.) (2000) *Audit cultures.* London: Routledge.

Swingewood, A. (1977) *The myth of mass culture.* London: Macmillan.

Tarr, J., Gonzalez-Polledo, E. and Cornish, F. (2018) 'On liveness: using arts workshops as a research method', *Qualitative Research*, 18(1), pp. 36–52.

Taylor, P. (2012) *Torkildsen's sport and leisure management.* Routledge. https://doi.org/10.4324/9780203877517.

Thiel, D. (2012) *Builders.* Abingdon: Routledge.

Tonnelat, S. (2010) The sociology of urban public spaces. *Territorial evolution and planning solution: experiences from China and France*, pp. 84–92.

Travel & Tourism Economic Impact | World Travel & Tourism Council (WTTC). (2021) Available at: https://wttc.org/Research/Economic-Impact (Accessed: 19 April 2021).

Tudor, A. (1999) *Decoding culture.* London: SAGE Publications Ltd.

Turner, B.S. (1998) *Status.* Milton Keynes: Open University Press.

Turner, B.S. (1999) 'McCitizens: risk, coolness and irony in contemporary politics', in Smart, B. (ed.) *Resisting McDonaldisation.* London: Sage. pp. 83–100.

Turner, B.S. and Rojek, C. (2001) *Society and culture.* London: SAGE Publications Ltd.

Turner, C. and Hodge, M.N. (1970) 'Occupations and professions', in Jackson, J.A. (ed.) *Professions and professionalisation.* Cambridge: Cambridge University Press.

Turow, J. (2005) 'Audience construction and culture production: marketing surveillance in the digital age', *The ANNALS of the American Academy of Political and Social Science*, 597(1), pp. 103–121. https://doi.org/10.1177/0002716204270469.

UEFA. (2016) *UEFA EURO 2016 – history – Ukraine-Poland – UEFA.com.* Available at: www.uefa.com/uefaeuro/season=2016/matches/round=2000448/match=2017898/prematch/background/index.html (Accessed: 21 May 2019).

University of Essex. (2021) *Welcome – home page – library services at university of Essex.* Available at: https://library.essex.ac.uk/home (Accessed: 23 July 2021).

Urry, J. (2002a) 'Mobility and proximity', *Sociology*, 36(2), pp. 255–274. https://doi.org/10.1177/0038038502036002002.

Urry, J. (2002b) *The tourist gaze.* 2nd edn. London: SAGE Publications Ltd.

Urry, J. and Larsen, J. (2011) *The tourist gaze 3.0.* 3rd edn. London: SAGE Publications Ltd.

Vaughan, D. (1996) *The challenger launch decision.* Chicago: The University of Chicago Press Ltd.

Veal, A.J. (2011) *Research methods for leisure & tourism.* 4th edn. Harlow: Pearson Education Ltd.

Visiting ExCeL. (2021) Available at: www.excel.london/ (Accessed: 15 August 2021).

Waheed, J. (2019) 'Nicki Minaj's concert in Slovakia is CANCELLED after arena suffers massive technical issues', *Daily Mail.* Available at: www.dailymail.co.uk/tvshowbiz/article-6736695/Nicki-Minajs-concert-Slovakia-CANCELLED-arena-suffers-massive-technical-issues.html (Accessed: 21 August 2021).

Ward, C. *et al.* (2010) 'Contextual influences on acculturation processes: the roles of family, community and society', *National Academy of Psychology (NAOP) India Psychological Studies*, 55(1), pp. 26–34. https://doi.org/10.1007/s12646-010-0003-8.

Warnaby, G. and Medway, D. (2013) 'What about the "place" in place marketing?', *Marketing Theory*, 13(3). https://doi.org/10.1177/1470593113492992.

Wearing, S., Stevenson, D. and Young, T. (2010) *Tourist cultures identity, place and the traveller.* 1st edn. London: SAGE Publications Ltd.

Wearing, S.L., McDonald, M. and Wearing, M. (2013) 'Consumer culture, the mobilisation of the narcissistic self and adolescent deviant leisure', *Leisure Studies*, 32(4), pp. 367–381. https://doi.org/10.1080/02614367.2012.668557.

Weber, M. (1978) *Economy and society.* London: University of California Press.

Welch, L.S., Benito, G.R.G. and Petersen, B. (2018) *Foreign operation methods.* 2nd edn. Cheltenham: Edward Elgar Publishing Ltd.

Wilks, J. and Page, S.J. (2003) *Managing tourist health and safety in the new millennium.* Oxford: Pergamon.

Williams, N.L., Ferdinand, N. and Bustard, J. (2019) *Tourism Review*, 75(1), pp. 314–318. https://doi.org/10.1108/TR-05-2019-0171.

Williams, R. (1988) *Keywords: a vocabulary of culture and society.* London: Fontana Press.

Wong, E.P.Y., Mistilis, N. and Dwyer, L. (2011) 'A framework for analyzing intergovernmental collaboration – the case of ASEAN tourism', *Tourism Management*, 32(2), pp. 367–376. https://doi.org/10.1016/j.tourman.2010.03.006.

Wood, S. (2018) 'SEC submits planning application for £150–200m "global event campus"', *Conference News.* Available at: www.conference-news.co.uk/news/sec-submits-planning-application-ps150-200m-global-event-campus (Accessed: 21 August 2021).

WOSP. (2020) Available at: www.wosp.org.pl/ (Accessed: 19 November 2020).

Wynn-Moylan, P. (2018) *Risk and hazard management for festivals and events.* 1st edn. Abingdon: Routledge.

Yeoman, I. (2012) *2050 – Tomorrow's tourism.* 1st edn. Bristol: Channel View Publications.

Young, M. and Muller, J. (eds.) (2014) *Knowledge, expertise and the professions.* Abingdon: Routledge.

Zamek-krolewski (2021) Available at: https://zamek-krolewski.pl (Accessed 9 March 2024).

Zaslavsky, V. and Brym, R.J. (1978) 'The functions of elections in the USSR', *Soviet Studies*, XXX(3), pp. 362–371. https://doi.org/10.1080/09668137808411193.

Zukin, S. (1991) *Landscapes of power: from Detroit to Disney World.* London: University of California Press.

Zukin, S. (2010) *Naked city: the death and life of authentic urban places.* Oxford: Open University Press.

Index